Palgrave Studies in Law, Neuroscience, and Human Behavior

Series Editors
Marc Jonathan Blitz, Law, Oklahoma City University School of Law, Oklahoma City, USA
Jan Christoph Bublitz, Faculty of Law, University of Hamburg, Hamburg, Germany
Jane Campbell Moriarty, Duquesne University School of Law, Pittsburgh, USA

Neuroscience is drawing increasing attention from lawyers, judges, and policy-makers because it both illuminates and questions the myriad assumptions that law makes about human thought and behavior. Additionally, the technologies used in neuroscience may provide lawyers with new forms of evidence that arguably require regulation. Thus, both the technology and applications of neuroscience involve serious questions implicating the fields of ethics, law, science, and policy. Simultaneously, developments in empirical psychology are shedding scientific light on the patterns of human thought and behavior that are implicated in the legal system. The Palgrave Series on Law, Neuroscience, and Human Behavior provides a platform for these emerging areas of scholarship.

Marc Jonathan Blitz

The Right to See with Technology

Recording, Augmented Perception, and the Constitution

Marc Jonathan Blitz
School of Law
Oklahoma City University
Oklahoma, OK, USA

ISSN 2946-5192   ISSN 2946-5206 (electronic)
Palgrave Studies in Law, Neuroscience, and Human Behavior
ISBN 978-3-031-89532-6   ISBN 978-3-031-89533-3 (eBook)
https://doi.org/10.1007/978-3-031-89533-3

© The Author(s), under exclusive licence to Springer Nature Switzerland AG 2025

This work is subject to copyright. All rights are solely and exclusively licensed by the Publisher, whether the whole or part of the material is concerned, specifically the rights of translation, reprinting, reuse of illustrations, recitation, broadcasting, reproduction on microfilms or in any other physical way, and transmission or information storage and retrieval, electronic adaptation, computer software, or by similar or dissimilar methodology now known or hereafter developed.
The use of general descriptive names, registered names, trademarks, service marks, etc. in this publication does not imply, even in the absence of a specific statement, that such names are exempt from the relevant protective laws and regulations and therefore free for general use.
The publisher, the authors and the editors are safe to assume that the advice and information in this book are believed to be true and accurate at the date of publication. Neither the publisher nor the authors or the editors give a warranty, expressed or implied, with respect to the material contained herein or for any errors or omissions that may have been made. The publisher remains neutral with regard to jurisdictional claims in published maps and institutional affiliations.

Cover illustration: VICTOR de SCHWANBERG/SCIENCE PHOTO LIBRARY/GettyImages

This Palgrave Macmillan imprint is published by the registered company Springer Nature Switzerland AG
The registered company address is: Gewerbestrasse 11, 6330 Cham, Switzerland

If disposing of this product, please recycle the paper.

Contents

1 **The Constitution, Enhanced Perception, and Freedom of Speech and Thought** 1
   The Constitution, Pictures, and Words 2
   Beyond Video Recording: Rights to See and Sense with Emerging Technologies 5
   References 24

2 **The First Amendment Right to Record and Seeing as "Speech Creation"** 27
   When a Right to Perceive Becomes a Right to Perceive with Technology 28
   Zemel v. Rusk, The First Amendment, and Our Lack of an "Unrestrained Right to Gather Information" 35
   The Rise of a Right to Record 41
   Rights to Record as a Right to Enhanced Seeing (and Thinking) 48
   References 55

3 **Freedom of Thought and Revisiting the Right to Receive Information (with Technology)** 57
   A Right to See Beyond "The Streetlight" 58
   The Right to Know—And the Right to Observe 60
   Placing Limits on a Freedom-of-Thought-Based Account of Perception 64

	Free Speech, Free Thought, and Government Purposes or Interests	66
	Social Practices and First Amendment Coverage (for Speech, Thought, and Perception)	81
	References	95
4	**The Right to Natural and Extended Vision (and Bodily and Mental Integrity)**	99
	The Right to See and Bodily Autonomy and Integrity	102
	Protection for Our Bodies—And Protection for Seeing That Comes with It	109
	Visual Processing, Brain, and Environment	112
	Extended Perception and Extended Mind	119
	The Right to Enhanced Perception in Unfamiliar Jurisprudential Territory	126
	A Summary of a Constitutional Framework for Perception Based on Personal Autonomy	132
	References	134
5	**Cognitive Liberty, Privacy, and Extended Perception**	137
	Moving Beyond "Core" Territories for a Right to See (in Expressive and Bodily Freedom)	138
	Rights, Interests, and Technological Change	140
	The Right to See and the Cognitive Construction of Our Perception	149
	Cognitive Liberty, Remote Sensing, and the Constitution	158
	Cognitive Liberty, Remote Seeing and Sensing, and Comprehensive Recording	161
	Privacy Threats and Deepfake Distortions	168
	Thinking About Rights, Technological Development, and Perceptual Enhancement: A Summary	179
	References	179

Index   183

CHAPTER 1

The Constitution, Enhanced Perception, and Freedom of Speech and Thought

Abstract In the early twenty-first century, American courts now agree that photography and video recording are protected by the First Amendment's protection for free speech. Image capture and recording, they conclude, constitute "speech creation." They enable individuals to share their experiences with others on social media. But courts and legal thinkers miss an important dimension of the rights-based protection we need for new technologies of seeing if they value such technologies only as means of communication. From lifelogging to telepresence to exploring virtual landscapes of fictional as well as real environments, technologies that augment our perception not only give us new ways of communicating—they give us new ways of reshaping our own minds and enlarging our experience—and to do so even when we don't serve as an audience for others' speech. As such, they are as much a component of freedom of thought as freedom of speech.

Keywords Freedom of speech · Freedom of thought · Right to record · Image capture · Speech creation · Virtual reality · Augmented reality · Extended reality · Lifelogging · Brain-computer interface device · Fourth Amendment · Surveillance

© The Author(s), under exclusive license to Springer Nature Switzerland AG 2025
M. J. Blitz, *The Right to See with Technology*, Palgrave Studies in Law, Neuroscience, and Human Behavior,
https://doi.org/10.1007/978-3-031-89533-3_1

The Constitution, Pictures, and Words

Pictures, including photographs and video images, are not only "worth a thousand words," as the proverb tells us. They are *treated* like words by American constitutional law. In the early twenty-first century, snapping a photograph or recording a video is—many courts have ruled—a form of First Amendment "speech."[1] Doing so is thus strongly protected against government restriction by the First Amendment's protection for speech.

This may at first seem puzzling. How does one engage in speech when one snaps a photograph or a video and doesn't say a word while doing so? Of course, videos and photographs can be mediums of expression: Photography and movie making are both integral to art and entertainment. But why would that make *every* kind of image capture—including photos that the photographer never plans to share—count as speech? Why would image capture be protected even when it lacks an artistic dimension?

The constitutional protection that courts have extended to image capture is less puzzling when one considers what image capture has become in the twenty-first century: An integral and pervasive part of modern communication.[2] A person clearly engages in First Amendment speech when they share an image or video on social media sites (such as YouTube, Facebook, or X). They do so even when they share it with a more limited audience in a text to a friend. Posting or sending a photo or video, after all, is an effective way of conveying information to one's audience. Or making that audience understand what it is to be immersed in a certain experience. A video or image is often far better than a verbal description at conveying the horrors of war or the destruction caused by a natural disaster.[3] Its vivid depiction of an event can often generate

---

[1] The Ninth Circuit recently said: "It defies common sense, to disaggregate the creation of the video from the video or audio recording itself. The act of recording is itself an inherently expressive activity." ALDF v. Wasden, 878 F.3d 1184, 1191 (2018).

[2] As Seth Kreimer has pointed out, "emerging technology and social practice have made captured images part of our cultural and political discourse." Seth F. Kreimer, *Pervasive Image Capture and the First Amendment: Memory, Discourse, and the Right to Record*, 159 U. Pa. L. Rev. 335, 373–74 (2011).

[3] *See* Margarita Alpuim and Katja Ehrenberg, *Why Images Are so Powerful—And What Matters When Choosing Them*, bonn institute, Aug. 3, 2023, at https://www.bonn-institute.org/en/news/psychology-in-journalism-5 (noting that "Images impact emotions faster and more powerfully than words").

more powerful emotion in its audience, and move that audience to action more effectively, than can any non-visual communication.[4] If any such sharing of images about matters of public concern were to be censored by the government, courts would almost certainly presume that such a restriction violates the First Amendment right to freedom of speech.

The sharing of video and images is also a pervasive part of modern life. For most of the twentieth century, taking and distributing photographs was a more burdensome and time-consuming way of sharing information than providing a verbal description. Apart from professional photographers and serious hobbyists, most individuals rarely carried a camera—doing so only on vacations or other special occasions—and the camera came with film that allowed for only a limited number of pictures. It took days or weeks to develop into photographs (usually by businesses specially dedicated to that function). Capturing *moving* pictures was even more burdensome and expensive. In the first decades of the twenty-first century, by contrast, powerful digital cameras are built into cell phones that people carry with them constantly and often need to participate in modern life. They can generate photographic images and videos instantaneously. Those who capture the images and videos can then send them to friends, or post them for audiences of thousands or millions, in a matter of seconds, and can store every picture they take in the massive memories of their computers or the servers they can access in the "cloud."[5] It is not

---

[4] As I noted in an earlier analysis of police body cameras, "video footage is far more able than eyewitness testimony to shift the debate from questions about what occurred in a police encounter to questions about how a just and well-functioning society should prevent excessive use of police force." See *Police-Worn Body Cameras: Evidentiary Benefits and Privacy Threats, American Constitution Society* (May 15, 2015), at https://www.acslaw.org/issue_brief/briefs-landing/police-body-worn-cameras-evidentiary-benefits-and-privacy-threats/#:~:text=Blitz%20admits%20that%20such%20cameras,police%20use%20of%20cameras%20in. *See* also Aaron Blake, *Why Eric Garner is the Turning Point Ferguson Never Was*, The Washington Post (Dec. 8, 2014); Michael Zhang, *The Power of the Camera in Recording and Shaping Social Movements*, PetaPixel (Jun. 13, 2020), at https://petapixel.com/2020/06/13/the-power-of-the-camera-in-recording-and-shaping-social-movements/.

[5] *See* Lee Davis and Rob Watts, *What Is Cloud Computing? The Ultimate Guide*, Forbes Advisor, Nov. 20, 2024, at https://www.forbes.com/advisor/business/what-is-cloud-computing/ (defining "the cloud" as "web-connected servers and software that users can access and use over the internet" allowing an individual to use applications without "having to host and manage [their] own hardware and software"—because the applications run from others' servers—and allowing them to use these applications anywhere they "have access to the internet").

surprising then that these visual records have become an integral part of the way individuals communicate in the twenty-first century.

Nor is it surprising that the First Amendment's constitutional shield—which is designed to protect communication of all kinds from censorship—protects communication with images. Or that it protects not only the sharing of an image or video but also its creation. The communicative practice we engage in when we convey information with videos is a multi-step process: Just as editorials in traditional newspapers have to be written and printed before they can be shared with an audience, videos have to be recorded. A censorious government could thus silence this kind of information-sharing not only by restricting the posting of a video at the moment it is shared but also by barring individuals from producing the video in the first place. As the Seventh Circuit Court of Appeals noted in *American Civil Liberties Union of Illinois v. Alvarez*, "laws enacted to control or suppress speech may operate at different points in the speech process . . . Restricting the use of an audio or audiovisual recording device suppresses speech just as effectively as restricting the dissemination of the resulting recording."[6] In fact, just as courts find that the First Amendment protects the writing of essays or poems when the writer ultimately chooses not to share that writing, American courts now generally agree that creating a video should count as First Amendment speech whether it is shared or kept private.[7]

Extending First Amendment free speech protection to video creation raises numerous questions: Do individuals have a right to record *everywhere* they have a right to be? Do they have a right to record *everything* they see there? If so, can the law protect the privacy of third parties whose actions and conversations might be recorded as these third parties read a book revealing their interests, talk with friends or family about confidential matters, visit a psychologist or other specialist, or associate with those who share their political views? If not, what limits exist on where, what, and how they can record? Do creators of videos have a right to enhance or augment their camera technology so it can capture details that would remain unseen if they relied on their natural vision and other capacities?

---

[6] ACLU of Ill. v. Alvarez, 679 F.3d 583, 595 (7th Cir. 2012). In stating that "[l]aws enacted to control or suppress speech may operate at different points in the speech process," the Seventh Circuit was quoting a Supreme Court case, Citizens United v. Federal Election Commission, 558 U.S. 310, 336 (2012).

[7] *See, e.g.*, Fields v. City of Philadelphia, 862 F.3d 353 (3d Cir. 2017).

For example, does the First Amendment include a right to record from an unmanned aerial vehicle (a "UAV" or "drone")?[8] Does it include a right to use high-powered magnification to reveal details invisible to the naked eye from their vantage point?[9] Or record large stretches or space over large periods of time, as map-making companies do?[10] These questions have all been the subject of past legal scholarship (including my own).[11] Some of them have been the focus of court cases—and will undoubtedly continue to be.

This book will briefly explore such questions—and more generally, examine how courts can address the apparent tension between the interest, protected by First Amendment law, in using camera technology to observe, record, and communicate about the world and individuals' interest in remaining *unseen and unrecorded*—so that they can maintain some of their privacy even as technologies of image and sound capture become more pervasive and powerful. It will also consider other interests that are threatened by modern video technology, including interests against being deceived by deepfake videos and other records that can make fictional events seem strikingly real.

Beyond Video Recording: Rights to See and Sense with Emerging Technologies

This, however, is not a complete picture of the way the United States Constitution should apply to new technologies of seeing and sensing. I will argue that there is something missing in the above speech-based account of our right to observe the world and capture records of what we perceive. In short, communication is only *one* of the things we can do with technologies that let us see and sense in new ways. A right to see or

---

[8] For an analysis of, and decision on this question, see Nat'l Press Photographers Ass'n v. McCraw, 90 F.4th 770, 777 (5th Cir. 2024). This issue is also discussed later in this book, in this chapter and Chapter 2.

[9] Dow Chem. Co. v. United States, 476 U.S. 227, 239 (1986).

[10] Marc Jonathan Blitz, *The Right to Map (and Avoid Being Mapped): Reconceiving First Amendment Protection for Information-Gathering in the Age of Google Earth*, 14 Colum. Sci. & Tech. L. Rev. 115 (2013).

[11] Marc Jonathan Blitz, James Grimsley, Stephen E. Henderson, Joseph Thai, *Regulating Drones Under the First and Fourth Amendments*, 57 Wm. & Mary L. Rev. 49, 89 (2015); Blitz, *The Right to Map, supra* note 10 (2013).

otherwise sense with technology is more than a right to *convey to others* the information that is in a picture we take or a film we shoot.

This becomes clearer as we expand our consideration of technologies of seeing *beyond* the example—familiar in this day and age—of recording a video with a smartphone camera and posting it on social media or sharing it with friends. Consider a few additional examples of augmenting perception with technology—many of which promise to bring into everyday life ways of seeing and recording the world which were once almost the exclusive province of science fiction. First, with the aid of "extended reality" technologies,[12] such as virtual reality ("VR")[13] and augmented reality ("AR")[14] we might not be confined to viewing a two-dimensional image or film of a past event or a far-away setting. We might instead be able to dive into such a scene, immerse ourselves in it, and perhaps interact with the people and objects we perceive there.[15] The story told by recorded light and sound isn't flattened onto a screen we merely *watch*. It is woven into an environment we are a *part of and can perhaps participate in*.[16]

---

[12] *See* U.S. Government Accountability Office, Science & Tech Spotlight: Extended Reality Technologies, Jan. 26, 2022, at https://www.gao.gov/products/gao-22-105541. (defining "extended reality" as "overarching term for a spectrum of technologies that link or integrate the digital world and the real world. These include augmented reality (AR), mixed reality (MR), and virtual reality (VR) technologies, all of which provide different degrees of sensory immersion and interaction between the real world and digital content").

[13] *See* Greg S. Weber, *The New Medium of Expression: Introducing Virtual Reality and Anticipating Copyright Issues*, 12 Computer/L.J. 175, 177 (1993) (stating that "[v]irtual reality is a computer-generated, three-dimensional world where participants have the illusion of walking or flying, manipulating objects, and interacting in real time" and giving people "[t]he feeling of immersion in a graphical environment and the ability to explore and navigate through this mysterious and floating atmosphere by using natural gestures, as if you were 'really there'").

[14] *See* Mark A. Lemley and Eugene Volokh, *Law, Virtual Reality, and Augmented Reality*, 166 U. Pa. L. Rev. 1051, 1054 (2018) (noting that while "VR replaces the real world altogether," "AR allows digital content to be layered over the real world").

[15] *See* Roya Bagheri, *Virtual Reality: The Real Life Consequences*, 17 U.C. Davis Bus. L.J. 101, 104 (2016) (noting that virtual reality technology "allow[s] a user to control movement and interaction within the virtual world").

[16] *See* Telepresence: Hearing Before the Subcomm. on Science, Technology and Space, 105th Cong. 14 (1998) (statement of S. Kicha Ganapathy, Member, Technical Staff, Multimedia Communications Research Laboratory, Bell Laboratories). The term, "telepresence," was suggested in an article by Marvin Minsky—based on a suggestion by futurist Pat Gunkel—to describe technology that allows one to work with tools that miles away by

Second, the light and sound that we view and listen to, whether in images on flat computer screens or in the virtual three-dimensional environments I have just described, might not come from a photograph or short video of the kind that we often see on social media—and that are at the heart of the "right to record" cases described earlier. It might instead come from far more intensive surveillance of people or events—surveillance that can last far longer or dig more deeply than the image capture that accompanies brief observation. This can include recording or observation with cameras that can see through walls (by "seeing" infrared radiation or other light invisible to the human eyes),[17] magnify small details from a distance,[18] or perhaps—with drones—from high in the air, or can last days or weeks, capturing not a few moments in the life of the person that is in the camera's field of view but rather long stretches of their day-to-day actions.

For much of the twentieth and early twenty-first centuries, such surveillance—when it could be conducted at all—was feasible on a large scale only for law enforcement officers and other government officials.[19] Courts asked when such intensive visual scrutiny or tracking of people by the state would count as a "search" that may be impermissible under the Fourth Amendment's privacy protections rather than permissible observation.[20] Now, the prevalence and affordability of such technology raise another constitutional question: When do the *rest of us*—private individuals who wish to use such powerful technologies for seeing, sensing,

---

feeling that they are in, or part of, our own hands, and where individual simultaneously as a 'sense of 'being there.'" *Marvin Minksy's Telepresnece Manifesto*, IEEE Spectrum, Aug. 31, 2010, at https://spectrum.ieee.org/telepresence-a-manifesto, reprinting Marvin Minsky, *Mind and Machine*, Omni, June 1980).

[17] *See* Michael A. De Vito and Stuart Flamen, *"Flir"ting With Danger: A Fourth Amendment Analysis of Infrared Imaging*, 10 Journal of Civil Rights and Economic Development (1995) (Infrared radiation is that part of the electromagnetic spectrum which is just beyond visible light, and thus undetectable by the naked eye").

[18] *See, e.g.*, Dow Chem. Co. v. United States, 476 U.S. 227, 239 (1986).

[19] *See* Marc Jonathan Blitz, *Video Surveillance and the Constitution of Public Space: Fitting the Fourth Amendment to a World that Tracks Image and Identity*, 82 Tex. L. Rev. 1349- 1481 (2004) (describing how, even in the early twenty-first century, video surveillance on a massive scale would be hard for any entity other than government to oversee).

[20] *See, e.g.*, Carpenter v. United States, 585 U.S. 296 (2018).

and capturing records of our surroundings—have *a right* to use these technologies to see and study the world around us?

When, for example, can we have a drone-mounted camera trail behind and hover over us, to record a jog through a natural setting or urban setting we find beautiful, or perhaps, a place that has great personal significance for us?[21] When can we wear glasses or visors, or a small camera on a shirt, that records every experience we have—allowing us to create a "lifelog" that lets us revisit any episode within a large stretch of our lives?[22] As other writers have pointed out, such a lifelog can serve many different purposes: It can, of course, powerfully supplement our otherwise unreliable and malleable memory of our own pasts. As Steve Mann, who pioneered the development of lifelogging technology, has said, it can serve as a "visual memory prosthetic."[23] It could also allow a person to better understand themselves and perhaps change their outlook or behavior—particularly if their lifelogging includes not only visual recordings of each act that they take but also information sensors that simultaneously gather about their physical and emotional states. It can provide a way for an individual to refute "deepfake" pictures or videos that can now show them doing or saying things they never do or said—and do so with vivid realism.[24]

These technologies for scrutinizing the world and generating visual or auditory records of it can, as I noted, not only produce images we can view on computer screens. They could conceivably be merged with the extended reality technology mentioned earlier to let us immerse ourselves in three-dimensional phantom copies of the places or scenarios our camera (or network of cameras) have captured. Rather than simply watching or

---

[21] *See*, Blitz, et al., *Regulating Drones, supra* note 11.

[22] *See* Anita L. Allen, *Dredging Up the Past: Lifelogging, Memory and Surveillance*, 75 U. Chi. L. Rev. 47, 49 (2008).

[23] Steve Mann, *Wearable Computing: A First Step Toward Personal Imaging*, Computer 30(2), Feb. 1997: 25–37, 27.

[24] Bobby Chesney and Danielle Citron have proposed that "immutable lifelogs or authentication trails" can be used to combat deception with deepfake videos in this way. *See* Bobby Chesney and Danielle Citron, *Deep Fakes: A Looming Challenge for Privacy, Democracy, and National Security*, 107 Cal. L. Rev. 1753, 1814 (2019). *See also* Steve Mann, *Continuous Lifelong Capture of Personal Experience with EyeTap*, CARPE'04: Proceedings of the 1st ACM workshop on Continuous archival and retrieval of personal experiences (2004) 2 (noting that a lifelog could enable a person to "prevent [] surveillance recordings [of that person] from being taken out of context.").

viewing a recording, we can experience it—and perhaps alter it as we do so—with artificial intelligence or other computer technology that can modify the tapestry of light and sound (and, in some VR technology, hardness we can feel as we touch a surface).

Science fiction writers of the past century have often imagined worlds where such remote sensing is part and parcel of everyday life. Various stories imagine VR technologies that give us "telepresence"—that is, the sense that we are present in an environment thousands of miles away. In Clifford Simak's 1944 story, *Huddling Place* (later a chapter in his book, *City*), he imagines a world where there is no need to physically travel anywhere because technology exists that allows us, with the turn of a knob, to virtually transport ourselves to a phantom recreation of someone else's office, so we can chat with them there, or to distant cultural or natural attractions and to what they have to offer thousands of miles away: concerts halls and library collections,[25] for example, or distant mountain ranges we can explore when we are actually sitting in our living rooms or offices.[26] You would not have to travel to a distant environment to see what is there because VR technology would instead bring a perfect simulation of that environment to you. As Simak wrote elsewhere, the same technology might instantly conjure up an invaluable training ground for testing skills of all kinds—allowing one to enter not only an environment that actually exists as a distant location, but a fantastical one. In his book, *Way Station*, the protagonist creates a "rifle range" that challenges him by confronting him with simulations of monstrous attackers.[27] Other writers have imagined worlds where people can use such technology to reexperience meaningful moments in their past with relatives who have died long ago.

We are now in a world where versions of this technology are available to individuals. In fact, innovative thinkers have been at work on such technology for decades. Virtual reality pioneers, such as Ivan Sutherland and Thomas Furness, began to develop computerized virtual reality displays in the 1960s.[28] Other VR developers, such as Frederick Brooks, Jaron

---

[25] Clifford D. Simak, *City* 41–60, Open Road Integrated Media (1980 [1952]); Clifford D. Simak, *Huddling Place*, Astounding Science Fiction (1944).

[26] *Id.*

[27] Clifford D. Simak, *Way Station*, Open Road Integrated Media (2015 [1963]).

[28] *See* Howard Rheingold, *Virtual Reality: The Revolutionary Technology of Computer-Generated Artificial Worlds—And How It Promises to Transform Society* 355 (First

Lanier, and Thomas Zimmerman, developed technology VR used can use touch and interact with the environments and objects within them.[29]

In the late 1970s and early 1980s, Steve Mann—the "father of wearable computers"—pioneered the development of wearable augmented reality devices, such as "eyetap" technology, that could "displa[y] computer information to the user and allo[w] the computer to process and augment what the user sees."[30] This technology, as Mann explained, can be used to create a comprehensive archive of personal experience—a lifelog, which he then used the technology to create and share.[31]

This could, he noted, provide each person with "an on-demand photographic memory" preserving "now-mundane details that might only later become important." The "mediated reality" made possible by such devices could also bring other benefits: Apart from allowing a user to extend their vision in both time (by recording and reviewing any moment of their life) and space (by watching a recording someone else has captured far-away), it could allow users to custom-design how they see in real time. As Mann pointed out, computer-mediated vision could highlight, and allow them to notice, parts of their environment they would otherwise likely go unseen; it could filter out numerous distractions (some of which, like advertising, are designed to capture and manipulate their attention).[32] Although these innovations often went further (even years ago) than many of the perceptual enhancement widely available to individuals today, some technology is beginning to make similar perceptual enhancing technology a more familiar part of life. Cameras on smartphones become more powerful each year, as do other small cameras that can be mounted on drones or vehicles. Consumers can purchase powerful lenses to expand the field of view or magnification power of many of

---

Touchstone ed. 1992) (1991); Fred Moody, *The Visionary Position: The Inside Story of the Digital Dreamers Who Are Making Virtual Reality* (1999).

[29] *See* Ken Pimental and Kevin Texeira, *Virtual Reality: Through the New Looking Glass* 42–45 (2d ed. 1994).

[30] *MannLab*, Eyetap, at https://mannlab.com/eyetap. *See also* Steve Mann, *Cyborg: Digital Destiny and Human Possibility in the Age of the Wearable Computer* (Doubleday of Canada 2001).

[31] *See generally* Mann, *supra*, note 24. *See also* International Electrotechnical Commission Blog, *The First Computer Wearable*, Feb. 26, 2024, at https://www.iec.ch/blog/first-computer-wearable.

[32] *See* Mann, *supra*, note 24.

these cameras. VR visors, such as Oculus Quest, HTC Vive, and, more recently, Apple's VisionPro,[33] allow people to see some recordings in three dimensions. Some visors can not only play recordings but allow individuals to make them. Or can function as augmented reality devices that display other relevant information (in the form of words or images) over the environment individuals view in front of them. Years after Mann's development of eyetap, Google, Meta, and other companies have sought to develop "smart glasses" that allow individuals to both record their surroundings and overlay them with augmented reality.[34]

The rise of brain-computer interface (BCI) technology and artificial retinas even raises the prospect that some capacities for enhanced seeing can even be incorporated into our bodies—rather than in cameras or visors outside it.[35] While the focus has been on developing such implantable technology to *restore* capacities that individuals have lost[36]— to restore sight to those who have become fully or partially blind, for example—that there is no reason that the same technology might not someday be used to enhance our vision, by building into it, for example, some of the same capacities for magnifying far-away sights or seeing normally invisible light that have already been incorporated into certain kinds of cameras.

It is conceivable that some of these technological enhancements of perception could fall under the same First Amendment protection that courts have found extends to video recording. It is also possible that some of them may fall within the territory of free speech protection for another reason: Someone donning an AR or VR visor, for example, might be engaging in First Amendment speech creation if they order a computer to draw map directions for them and superimpose it over the tangle of streets that lies in front of them. Or if they ask it to recreate the image

---

[33] *See* Rhodri Marsden, *I Tried the Apple Vision Pro headset. Did I See the Future?* Financial Times, Oct. 7, 2024, at https://www.ft.com/content/387f3293-01f1-4e95-b226-2db93c00b52d.

[34] *See* Aamir Siddiqui, *Meta's First AR Glasses Are a Spiritual Successor to Google Glass, but You Can't Buy Them*, Android Authority, Sep. 26, 2024.

[35] Jeramiah D. Wander and Rajesh P.N. Rao, *Brain–Computer Interfaces: A Powerful Tool for Scientific Inquiry.* 25 Curr Opin Neurobiol. Apr, 2014.

[36] *See* R. A. Juskalian, *New Implant for Blind People Jacks Directly into the Brain*, MIT Technology Review, Feb. 6, 2020, at https://www.technologyreview.com/2020/02/06/844908/a-new-implant-for-blind-people-jacks-directly-into-the-brain/.

of a long-razed historical building on a site within a city that it stood in decades ago. Electronically conjuring such a phantom building or landscape, in order to observe and explore it, is arguably a technologically advanced form of creating or viewing a painting or film, two activities that have for decades been recognized by Americans to be forms of expression protected by the First Amendment.[37]

However, at least some of what I have just described cannot easily be described as "speech creation" or receipt of others' speech. Using a bionic eye or artificial ear to restore lost visual or hearing capacity may be invaluable for an individual. So, in some cases, might be use of a bionic eye or another artificial sight-generating device that can give its user night vision or the capacity to see directly in back of them even as they are facing forward, or to use computer-mediation of vision to filter out content they don't wish to see or enhance their perception of what they do want to see. But restoring or enhancing vision in this way isn't speaking. Nor is it clear how viewing images on the live feed of a surveillance drone or street camera constitute speech. It is true that viewing such a live feed may in some cases be a *precondition* to accurately writing about or otherwise describing them. But racing a car or climbing a mountain is similarly a precondition to describing such experiences and that does not transform car racing or mountain climbing into First Amendment speech.

One conclusion one might draw from these reflections is that some enhancements of our perception are simply unprotected by the Constitution because they lie outside the coverage of the only right that might protect them. In this book, however, I want to explore another possibility: We have a right to see and sense *with technology*, one might argue, because that is a consequence of the fact we have a right to see and sense *without it*. We have a right to view and listen to the world with our *unaided* eyes and ears. Imagine the government developed technologies that could block us from seeing or hearing what is in our immediate surroundings. Or that it could control or distort our vision or hearing by generating hallucinations in our heads. This wouldn't be censorship of the kind traditionally barred by the First Amendment. In some cases, perhaps, it would entail physical invasions of the kind that violate a different constitutional right in the American system: The right to bodily integrity and autonomy

---

[37] *See* Mann, *supra*, note 24 (noting that technology for creating a lifelog would be "more than a visual memory prosthetic" and would also provide a "a new tool for the visual arts.").

that individuals receive under the Fifth and Fourteenth Amendments. If officials forcibly blindfolded a person, for example, or used a chemical spray to disable that person's vision, they would in doing so be intruding into that person's constitutionally protected realm of bodily autonomy and physical liberty. But even if officials could interfere with our natural sensory processes from a distance, or perhaps, interfere not with our receipt of sensory stimuli but rather with the way our brain processes such stimuli, their doing so would unquestionably violate a constitutional right individuals have to be free of coercive intrusions into their person. And if government is prohibited by the Constitution from engaging in such interference by interfering in our natural perceptual processes why not also *technological extensions* of those processes that individuals in modern societies come to rely upon? Imagine, for example, that government officials left a person unable to see clearly not by interfering in the working of that person's eyes or brain processing but rather by seizing (or making it illegal) for them to wear glasses or use an augmented reality technology designed to compensate for some deficit in the way their brain processes information. Or that it did so by *compelling* individuals to implant bionic eyes or wear augmented reality visors the operation and outputs of which were under the control of government officials. Or by surreptitiously hijacking a visual prosthetic someone regularly uses—and perhaps *needs* to use to see well—so that it shows that person not what they wish to see but rather what the government wants them to see.

First Amendment free speech rights might here provide guidance in understanding a right to see or sense with technology—but that is *not* because the latter right is necessarily an instance of the former but rather because it is analogous to it: The protection the Constitution provides for freedom of speech doesn't only protect us when we express ourselves with our vocal cords or a pen and paper but also when we do so with modern technology—when we write a paper on a computer word processing program, send a text message or social media post, include a message in a web site, or use amplification equipment to project our voice to thousands of people at a rally. Or, as noted earlier, when we convey information by sharing photographs or videos. The government would run afoul of the First Amendment's free speech protection not only by forbidding us from expressing our thoughts in words but also by prohibiting use of the technologies necessary for us to communicate in the twenty-first century—and reach the general public. Just as we have a right not only to speak but to speak with technology so we may have a right not only to see but to see with technology. Just as our right to freedom of speech—in the twenty-first century—covers use of technologies that allow us to speak to large audiences and not simply individuals directly around us so, one might argue, whatever right we have to perceive our environment should embrace not only natural capacities or tools we might use to see what is immediately in front of, or beside us, but also to observe (in at least some manner) other parts of our public environment and to do so in ways that go beyond the seeing and sensing that was possible in a less technologically-advanced age.

To be sure, one might make an argument that the above addendum to judicial protection of a right to record should be understood as a modest one: Our right to the integrity and autonomy of our person, shields our biological sensory processes from government interference. Perhaps it also shields certain other technologies we wear or carry on our person, such as glasses or hearing aids, to restore lost capacities. But it does *not* give us a wide-ranging right to use all manner of surveillance tools or other technologies of seeing *except* to the extent that such remote observation comes to us, via First Amendment speech (such as a recording), from someone who has a right to create it. This is an account of our right to use enhanced perception that I will consider more closely in Chapter 2 (on the right to record), in Chapter 4 (on protection of our visual system), and in the first part of Chapter 5 (continuing the discussion of both of those).

There is, however, another key feature that this book will argue underlies the right to see and sense with technology and that, in some respects, weighs in favor of defining it more broadly. Such a right is not merely a right to observe the environment. It is *also* a component of a right to "freedom of thought"[38] or "freedom of mind."[39] Consider again the enhancements of perception I have described above: Lifelogging, the experience of walking on a hallucinatory streetscape or other imaginary landscape, or having telepresence in a real location far away, whether to simply enjoy our presence there, to learn about the structures and artifacts that surround us, or to retrain our minds by reshaping the way we respond to encounters we have there. The value we find in such

---

[38] *See* Jones v. Opelika, 316 U.S. 584, 595; Jones v. Opelika, 316 U.S. 584, 616, 618 (1942) (Murphy, J., dissenting).

[39] *See* Wooley v. Maynard, 430 U.S. 705, 714 (1977) (quoting W. Va. State Bd. of Educ. v. Barnette, 319 U.S. 624, 637 (1943)).

perceptual experience doesn't generally consist only of passively absorbing information from our environment (as discussed in Chapter 5, even that is not as passive as it seems, since our mental processing is always actively constructing our conscious percepts). Rather, it is taking steps to *consciously* give form to what our minds experience or consciously shape (reshape) the way our minds work. The interest we have in using such technologies then lies in exercising a kind of mental self-determination or "cognitive liberty."[40] A right to see or sense with technology is a component of that general right to "freedom of mind." It is not merely a component of a right to see or receive information we might later share. As a right that enables us to exercise our mental powers in certain ways, it may require more—in some circumstances—than being able to see what is in our immediate environment.

As a consequence, we should situate the legal analysis of these technologies in larger debates about how to protect our cognitive liberty—and define its limits—in the first decades of the twenty-first century. The right to see and sense with technology is, at least in part, a component of what I have previously described as "a right to think with technology."[41] It is a right to exercise and reconfigure our mental powers not simply with the natural capacities we have been given to think with but with the tools that social and technological development supplies for us.

Past scholarship on the right to record information has already stressed its link to freedom of thought. In an article on the constitutional status of "pervasive image capture," for example, Seth Kreimer notes that photographic images are not only a key component of twenty first-century discourse—they are a way of augmenting and strengthening our natural memory. While "visual memory" is notoriously "thin and unreliable," camera records don't fade or alter their shape with time: They allow us to re-perceive the event they captured.[42] Justin Marceau and Alan Chen similarly stress that images we capture or videos we record with phone

---

[40] I will discuss the term "cognitive liberty" more fully in Chapter 5. In a journal article called "On Cognitive Liberty", Richard Glen Boire that "the right to control one's own consciousness is the quintessence of freedom" and that exercising this right is not only a matter of exercising free speech but also using other tools to share one's conscious. *On Cognitive Liberty*, Part I, Journal of Cognitive Liberties 1(1): 1–3 (1999–2000).

[41] Marc Jonathan Blitz, *Artificial Minds in First Amendment Borderlands* in ed. Woodrow Barfield, *The Law of Algorithms* 604 (Cambridge University Press 2020).

[42] Kreimer, *supra*, note 2, at 342.

cameras (or other cameras) can not only provide raw material for communications with others, but also serve as more vivid functional equivalents of notes we write for ourselves—to memorialize our own observations of an event.[43] Our use of language in a diary or private journal entry is less an exercise of communication than an exercise of our freedom to think (and remember) and so, on these accounts, is some photography and video creation. Jane Bambauer likewise argues that the First Amendment's commitment to the protection of "free thought and information flow" requires it to protect our ability not only to create and share images but also to generate and exchange other kinds of data. Regulations that thwart that, she argues, interfere with our "liberty interest in knowledge creation."[44] My past work argues that protecting freedom of thought may require protecting not just natural observation and video recording but also technological enhancements to such recording and the way we access it (for example, in high-tech global maps): "[O]ur ability to think freely," it has argued, "depends not only on our ability to communicate free of government monitors and censors, but also to observe the world free from government-imposed blinders." Such freedom, I argued, should include a right not only to perceive our immediate environment or to receive recordings but also to use VR and other computer technology to enable immersive and "virtually direct perception of far-away environments."[45]

In a sense, the arguments in this book build on all of these previous analyses. However, in doing so, they draw a sharper distinction between the First Amendment doctrine that has so far evolved to shield recording and other image capture or recording of our perceptions and a right to see or sense rooted in cognitive liberty. In doing so, it will discuss and elaborate a key reason to avoid conceiving of our right to see with technology as merely a component of a right to speak with it: To do so transforms recipients of a recording or live video feeds into an audience that is *dependent*, in key respects, on the choices made by *someone else*, namely the person whose "speech" they receive (in the form of a recording or video transmission). Video recording, as noted above, allows for a kind of

---

[43] Alan K. Chen and Justin Marceau, *Truth and Transparency: Undercover Investigations in the Twenty-First Century* 162, 165 (Cambridge University Press 2023).

[44] Jane Bambauer, *Is Data Speech?*, 66 Stan. L. Rev. 57, 83 (2014).

[45] *See* Blitz, *Right to Map*, *supra*, note 10, 82, 122–123.

extended seeing—allowing the perceiver to see something distant in space or time. But viewing such a video on social media or a text is unlike the use of our natural vision, which we can use to scan and learn about our immediate environment whenever we would like to do so. Just as a reader can't read a book until an author writes it, and a listener can't hear someone's verbal statement until the speaker makes it, we can't view a video we rely upon to directly see a certain location until it is recorded and posted by its creator (and perhaps by others who play a role in distributing it on social media or other platforms). This extension of our seeing—with visual records created by others—has tremendous value. Many recordings that inform our thinking about democracy take this form.

Yet there are also forms of extended seeing that allow the perceiver to be more *self-reliant* and play a more active role in selecting what to observe and scrutinize. Think again about the technologies—discussed above—for extending seeing with technologies of extended reality (or "XR"). With a combination of cameras, computers, and VR technology such as a specialized headset, I can gain "telepresence" in a distant location. I can feel as though I am physically there.[46] In *this* kind of extended seeing, I may be able to explore a virtual environment on my own initiative—the way I can choose what to observe in the physical space that surrounds me.[47] Depending on the technological apparatus that endows me with such telepresence, I may even be able to change my vantage point—for example, by virtually walking (or, perhaps, teleporting) a few blocks down a street in a distant city to see a different part of it, or to see some of the same structures from a different perspective. In contrast to a posted video recording, where I can only see what the camera-wielder has chosen to film, with this kind of technology, I may be able to decide for myself what to focus on and where to see it from. As Mann notes, wearable computing could also allow us to exercise more fine-grained control over how and what we see—to alter our visual reality by mediating it.[48]

Other less immersive technologies may also allow for forms of extending seeing or sensing in which I act with more autonomy than I do when I am an audience of others' visual records. Imagine, for

---

[46] *See Telepresence: Hearing Before the Subcomm. on Science, Technology and Space*, 105th Cong. 14 (1998), *supra* note 16.

[47] *See* Tony Reichhardt, *Almost Like Being There*, Smithsonian Magazine, Dec. 2015.

[48] *See* Mann, *supra*, note 24.

example, that I view two-dimensional images on a computer or mobile device screen rather than interacting with three-dimensional immersive VR environments: In some cases, these flat images might come to me not in the form of others' recordings, but from a camera or camera that I operate myself—perhaps on a drone that I pilot. I may also be given greater control over what I see even if I rely on *someone else's* camera apparatus to observe a distant location—if that apparatus allows me (and presumably other users) to select from, and switch between, multiple different vantage points. In fact, this can be true even if my window on the world is provided not with a live camera feed but rather by a computer program that lets me navigate through a simulacrum of the world built from millions of photographs captured by planes, trucks, and other vehicles. This, for example, is what occurs when a person explores the world with Google Streetview or other "virtual globes."[49] Streetview can give me the sense of visually gliding down a boulevard in a far-away city or hiking a trail near the Grand Canyon—and perhaps also flying over a cityscape—and it does so *not* by giving me access to a live video feed from those locations but *rather* by letting me navigate through a vast compilation of images that have captured such environments comprehensively enough that movement from one image to another feels like travel within the environment.[50] It can also let me "zoom in" on such locations—by showing me images that appear to bring a particular building, street, or other structure closer and closer to my eyes. Just as a vast library of books gives a reader a *choice* over what ideas they will explore or stories they will enter into—or a kind they do not have when they can only read a single book presenting a single author's experience or perspective, so the vast library of images assembled by virtual globe systems gives individuals a level of control over their observations that they don't have when they are confined to the four corners of a single video recording shared on social media.[51]

---

[49] *See* C.D. Elvidge and B.T. Tuttle, *How Virtual Globes Are Revolutionizing Earth Observation Data Access and Integration* The Int'l Archives of the Photogrammetry, Remote Sensing and Spatial Information Sciences. Vol. XXXVII. Part B6a. Beijing (2008), at 138 ("Virtual globes have democratized the access to global satellite imagery. Anyone can download the basic software for free and have open access to a full global mosaic of earth observation satellite imagery").

[50] *See* Blitz, *Right to Map*, *supra* note 10.

[51] *See* id. at 186. *See also* Anne Klinefelter, *Reader Privacy in Digital Library Collaborations: Signs of Commitment, Opportunities for Improvement*, 13 I/S: J.L. & Pol'y

The latter lets me see only the world the video recorder has captured as they have captured it. It is true I may be able to notice details in a recording that the camera-wielder missed. I may be able to pause and focus on aspects of the recording they didn't attend to when they made it. I might also watch a hundred times a sequence of recorded events that they perhaps only saw once (in making the recording). There is a place, in other words, for active scrutiny and analysis even when I act merely as an audience for someone else's completed video recording. Yet, there are also respects in which I cannot place myself "behind" the recording and gather more information about the reality it depicts. I can't rewind the recording to watch events that occurred before it starts or visually observe the event after the recording ends. Nor can I switch the angle from which the video is made or change the camera lens it is captured by (at least in the kinds of recordings that we typically find on social media or other platforms in the present age). I am confined, in watching such a recording, to the boundaries of the visual representation someone else created—boundaries I may be able to move beyond in *other* kinds of seeing with technology. As Mann notes, wearable computer can even allow me to exercise more fine-grained control over how and what I see as I see it: it could allow me to select and switch between one "visual reality" and another by changing how the computer processes the light that reaches me.[52]

In fact, I may be able to exceed such boundaries—to exert greater control over what I see and from where—not only with the "hi-tech" XR technologies, drone mounted cameras, or virtual globes described earlier but also with more familiar technology. Imagine, for example, that I wish to observe the skies with a powerful telescope from my own property or that of a friend, or from a public park. Or that I want to use binoculars to observe plants, animals, or geological features in a nature reserve. When I do so, I supplement my natural vision with magnification technology and can make my own choices about what to observe.

I can also see in a way that is, in a sense, more *solitary* than being part of a communication that involves sharing of video or audio recordings. The light that reaches my eyes through a telescope or binoculars comes

for Info. Soc'y 199, 200, 208 (2016) (explaining that "[p]rivate exploration of ideas is defended as a precondition to autonomy" and how library collections can contribute to such exploration).

[52] See Mann, *supra*, note 23, at 27–28.

to me not through another person's efforts to convey certain information to me but rather directly from the natural world. The same may be true when I watch a video feed sent from a drone camera or obtain telepresence in an immersive replica of a distant landscape. In all of these cases, I will often be using technology designed by others. In some, I will be using technology that is—even as I use it—owned and managed by others: If I use Google Streetview to traverse the streets of a distant city, for example, I am using software and data that has not only been compiled by Google but also constantly managed and updated by it.[53] In all of these cases, however, I am perceiving the world using technologies that allow me to explore it myself—and to do so with the mediation of a machine (for example, a camera I control) rather than another person. Justice Robert Jackson wrote in 1943 that First Amendment expression establishes a "short cut from mind to mind"[54] and video recordings as well as words can serve as such a shortcut, by letting me see what others have seen and recorded. However, the different forms of extended seeing I am describing here instead establish another kind of shortcut—one not from another mind to my own but rather from the *external world* to my mind—and that is true even when that shortcut is lengthened to some extent by the complex intermediation of cameras, computers, internet transmissions, and other mechanical processes. When using this kind of technology, I don't need another person to share *their* view of the world with me, because machines allow me to view it myself—perhaps with certain social practices that allow me to use them this way.

Many technologies of extended seeing, then, can free a perceiver from the constraints they are under when they are simply an audience for the video recordings shared with them by others—and they can do so in at least two ways: By (1) giving a perceiver greater control over how what they can see and focus on and from what vantage point (and perhaps over how the technology mediates what they see) and (2) by letting the perceiver observe an environment, and perhaps interact with it, in a solitary interaction with their environment—one in which they don't need to trust in anyone else's description of it, or anyone else's choices about how to create a record of it. When they use a powerful drone camera to

---

[53] *See* Mark Davis, *How Does Google Earth Work?*, LiveScience, May 17, 2019, at https://www.livescience.com/65504-google-earth.html.

[54] W. Virginia State Bd. of Educ. v. Barnette, 319 U.S. 624, 632 (1943).

observe a landscape, for example, they can choose where to position the drone, where to point the camera mounted on it, and whether to magnify or otherwise enhance the image. They can also do so without needing a First Amendment *speaker* to serve as an intermediary: It is the mechanical processes in the drone and camera that provide a conduit between the information in the drone's environment and the person watching it.

Why is this significant for understanding when and how the Constitution shields our use of technologies to perceive the world around us? Because, as I have suggested earlier and will argue more fully in Chapter 2 and other chapters of this book, the solitary exploration and learning we do when we exercise our power of perception is not best conceptualized as invariably or primarily part of our First Amendment freedom of communication. After all, we don't necessarily communicate with anyone when we view the world around us or create a record to privately view later. We are rather exercising another component of our intellectual or mental liberty. Years ago, the Supreme Court described "the freedom to speak" as only one part of a broader "freedom of mind."[55] In some cases, perhaps, use of technologies for perception—or creating records of what we perceive—falls squarely within the former freedom: When we convey a video or audio recording or image to another person we are, in a sense, speaking with it. At other times, however, the way see with technology falls *outside* of this "freedom to speak" and constitutes another, *different* kind of exercise of our freedom of mind. That is what this book shall argue.

Moreover, a right to perceptual enhancement of this kind might help address another challenge. As I mentioned earlier, in past decades it is only government officials that have been able to make large scale use of powerful technologies of surveillance to collect information about individuals. A right to perceive the world with technology, has other writers have observed, allows citizens to do so as well—and counter any state monopoly on information gathering and to engage in observation of state observers. Steve Mann, Jason Nolan, and Barry Wellman use the term "sousveillance" to describe such use of video or other technologically-enabled monitoring (and countering) of surveillance. "Sousveillance,"

---

[55] *Wooley*, 430 U.S. at 714.

they write, "focuses on enhancing the ability of people to access and collect data about their surveillance and to neutralize surveillance."[56]

Chapters 2 and 3 start by considering arguments for grounding a right to record—and perhaps other kinds of technological enhancements to perception—in First Amendment protection for "speech creation" or a "right to receive information and ideas." Chapter 2 focuses on the doctrine that started this Chapter: First Amendment protection for communication and expression. It looks at how courts have analyzed the right to record and how they (or legal scholars) might try to adapt this framework to other ways of seeing or sensing with technology. Ultimately, I will argue, a traditional First Amendment framework focused on speech can only help us so much. Courts and legal scholars need to go *beyond* it by recognizing that protecting the way we say things with images doesn't fully protect the way we see and think with them.

Chapter 3 then considers the possibility that we might locate a right to see within a broader First Amendment protection for creation of, and access to, information. The Supreme Court has, since 1943, recognized that when the Constitution protects the freedom of speech, it protects not only giving voice to it but also acting as an audience for it. It protects readers, listeners, and viewers as well as speakers. But all of these are audiences for *speech* of some kind (an oral statement, perhaps, or a book, or a painting or movie). Those whose information comes from their observation of the world itself have faced more skepticism in claiming First Amendment protection for their observation. The Constitution, the Supreme Court has emphasized, contains no right to be free of restriction whenever one gathers information by learning something about the world—because every act we might take, even non-speech acts that come with significant risks of economic and physical harm, potentially teach us something. Being in a fistfight, for example, can teach us what that experience is like—and perhaps how to be a more effective fighter. Still, scholars have offered answers to that question, answers that explain how courts might protect our capacity to observe our environment *without* disrupting it (and threatening others' interests as we do so)—or give us a right to *know* what is in data without giving us a right to insulate any harmful or dangerous activity taken to generate it. Chapter 3 will

---

[56] *See* Steve Mann, Jason Nolan, and Barry Wellman, *Sousveillance: Inventing and Using Wearable Computing Devices for Data Collection in Surveillance Environments*, Surveillance & Society 1(3): 331–355 (2002).

discuss such answers and acknowledge their power but also argue that this approach too at best provides only a partial foundation for a right to see with technology that protects the underlying interests at stake.

Chapters 4 and 5 then turn to less familiar territory in discussions of how the Constitution might protect our perceptions, and technologies that allow for them. Chapter 4 goes beyond the First Amendment analysis in which we conceive of the right to see as a component of, or antecedent to, a right to communicate. It asks if we can instead conceive of it (as I noted we might do above) as a part and parcel of our right to bodily or personal integrity. It explains why this approach is an important addition to the analysis of a right to observe the environment around us: We have a right to view the world with our *natural* vision in part because government measures that prevent us from doing so would almost inevitably interfere with our physical freedom and the decisions we make about what to do with our own bodies and how to interact with the environment that directly surrounds them. Matters become more complex when we consider the way we use technologies that lie outside of our bodies—from binoculars and telescopes to cameras and computers.

Chapter 4 then analyzes these questions by turning to certain arguments about neuroethics and the ways law should protect our mental freedom by protecting the integrity of our brain and nervous system. It does so by looking for guidance to ethical and legal theories built around the concept of "the extended mind"—as well as other analysis of what tools that lie outside of our person can be considered part and parcel of our personal autonomy. It also asks how a variant of such a framework might apply to technologies that extend our perception—and some of the difficulties raised by the fact that, as available as some perception-enhancing technologies are, they can also raise harms for the perceivers themselves or for those they observe or otherwise carry information about.

Chapter 5 then sketches some tentative answers to the concerns that conclude Chapter 4. It takes a closer look at why protection for a right to see with technology might extend beyond simply protecting artificial eyes or brain-computer interfaces we use to replace lost visual function. Some technologies that go beyond such restoration of our sensing powers may be necessary to let our powers of perception further interests we need them to further—interests that our natural unaided vision may not be able to further without technological augmentation.

Chapter 5 also explores two kinds of interest that might, in some circumstances, cut against a right to see with technology—and demand limits on it in the form of what I have called a "counter right" or a basis for "counter-coverage": (1) A right to remain unseen, to preserve privacy, and (2) a right against distortion of what we or others see. In this concluding chapter, I will summarize my argument: Given many new emerging technologies for augmenting or generating our perceptions, the law needs to provide a foundation for a right to see with technology. It should root this right in a larger right to *think* with technology. Far from simply being derivative of a right to see, a right to perceive the world should be understood as a component of a larger right to freedom of thought and intellectual liberty.

References

Articles and Books

Anita L. Allen, *Dredging Up the Past: Lifelogging, Memory and Surveillance*, 75 U. Chi. L. Rev. 47 (2008).

Margarita Alpuim and Katja Ehrenberg, *Why Images Are so Powerful—And What Matters When Choosing Them*, bonn institute, Aug. 3, 2023, at https://www.bonn-institute.org/en/news/psychology-in-journalism-5.

Roya Bagheri, *Virtual Reality: The Real Life Consequences*, 17 U.C. Davis Bus. L.J. 101, 104 (2016).

Aaron Blake, *Why Eric Garner Is the Turning Point Ferguson Never Was*, The Washington Post (Dec. 8, 2014).

Marc Jonathan Blitz, *Video Surveillance and the Constitution of Public Space: Fitting the Fourth Amendment to a World That Tracks Image and Identity*, 82 Tex. L. Rev. 1349 (2004).

Marc Jonathan Blitz, *The Right to Map (and Avoid Being Mapped): Reconceiving First Amendment Protection for Information-Gathering in the Age of Google Earth*, 14 Colum. Sci. & Tech. L. Rev. 115 (2013).

Marc Jonathan Blitz, Marc Jonathan Blitz, James Grimsley, Stephen E. Henderson, Joseph Thai, *Regulating Drones Under the First and Fourth Amendments*, 57 Wm. & Mary L. Rev. 49, 89 (2015).

Marc Jonathan Blitz, *Artificial Minds in First Amendment Borderlands* in ed. Woodrow Barfield, The Law of Algorithms. Cambridge University Press (2020).

Richard Glen Boire, *On Cognitive Liberty*, Part I, Journal of Cognitive Liberties 1(1): 1–3 (2000).

Mark Davis, *How Does Google Earth Work?*, LiveScience, May 17, 2019, at https://www.livescience.com/65504-google-earth.html.

Lee Davis and Rob Watts, *What Is Cloud Computing? The Ultimate Guide*, Forbes Magazine, Nov. 20, 2024.

Michael A. De Vito and Stuart Flamen, *"Flir"ting With Danger: A Fourth Amendment Analysis of Infrared Imaging*, 10 Journal of Civil Rights and Economic Development (1995).

C.D. Elvidge and B.T. Tuttle, *How Virtual Globes Are Revolutionizing Earth Observation Data Access and Integration*, The Int'l Archives of the Photogrammetry, Remote Sensing and Spatial Information Sciences. Vol. XXXVII. Part B6a. Beijing (2008).

R. A. Juskalian, *New Implant for Blind People Jacks Directly into the Brain*, MIT Technology Review, Feb. 6, 2020, at https://www.technologyreview.com/2020/02/06/844908/a-new-implant-for-blind-people-jacks-directly-into-the-brain/.

Anne Klinefelter, *Reader Privacy in Digital Library Collaborations: Signs of Commitment, Opportunities for Improvement*, 13 I/S: J.L. & Pol'y for Info. Soc'y 199, 200 (2016).

Seth F. Kreimer, *Pervasive Image Capture and the First Amendment: Memory, Discourse, and the Right to Record*, 159 U. Pa. L. Rev. 335, 373–74 (2011).

Mark A. Lemley and Eugene Volokh, *Law, Virtual Reality, and Augmented Reality*, 166 U. Pa. L. Rev. 1051, 1054 (2018).

Steve Mann, *Wearable Computing: A First Step Toward Personal Imaging*, Computer 30(2): 25–37 (Feb. 1997).

Steve Mann, Jason Nolan, and Barry Wellman, *Sousveillance: Inventing and Using Wearable Computing Devices for Data Collection in Surveillance Environments*, Surveillance & Society 1(3): 331–355 (2002).

Rhodri Marsden, *I Tried the Apple Vision Pro headset. Did I See the Future?* Financial Times, Oct. 7, 2024, at https://www.ft.com/content/387f3293-01f1-4e95-b226-2db93c00b52d.

Marvin Minksy's Telepresence Manifesto, IEEE Spectrum, Aug. 31, 2010, at https://spectrum.ieee.org/telepresence-a-manifesto, reprinting Marvin Minsky, *Mind and Machine*, Omni, June 1980).

Fred Moody, *The Visionary Position: The Inside Story of the Digital Dreamers Who Are Making Virtual Reality* (1999).

Ken Pimental & Kevin Texeira, *Virtual Reality: Through the New Looking Glass* 42–45 (2d ed. 1994).

Tony Reichardt, *Almost Like Being There*, Smithsonian Magazine, Dec. 2015.

Howard Rheingold, *Virtual Reality: The Revolutionary Technology of Computer-Generated Artificial Worlds—And How It Promises to Transform Society* 355 (First Touchstone ed. 1992) (1991).

Aamir Siddiqui, *Meta's First AR Glasses Are a Spiritual Successor to Google Glass, but You Can't Buy Them*, Android Authority, Sep. 26, 2024.

Clifford D. Simak, *City*, Open Road Integrated Media (1980 [1952]); Clifford D. Simak, *Huddling Place*, Astounding Science Fiction (1944).

Clifford D. Simak, *Way Station*, Open Road Integrated Media (2015 [1963]).

Jeramiah D. Wander and Rajesh P. N. Rao, 25 Brain–Computer Interfaces: A Powerful Tool for Scientific Inquiry. Curr Opin Neurobiol. Apr, 2014.

Greg S. Weber, *The New Medium of Expression: Introducing Virtual Reality and Anticipating Copyright Issues*, 12 Computer/L.J. 175, 177 (1993).

Michael Zhang, *The Power of the Camera in Recording and Shaping Social Movements*, PetaPixel (Jun. 13, 2020).

Government Reports

Telepresence: Hearing Before the Subcomm. on Science, Technology and Space, 105th Cong. 14 (1998) (statement of S. Kicha Ganapathy, Member, Technical Staff, Multimedia Communications Research Laboratory, Bell Laboratories).

U.S. Government Accountability Office, Science & Tech Spotlight: Extended Reality Technologies, Jan. 26, 2022, at https://www.gao.gov/products/gao-22-105541.

Cases and Statutes

ALDF v. Wasden, 878 F.3d 1184 (2018).

ACLU of Ill. v. Alvarez, 679 F.3d 583, 595 (7th Cir. 2012).

Carpenter v. United States, 585 U.S. 296 (2018).

Dow Chem. Co. v. United States, 476 U.S. 227, 239 (1986).

Fields v. City of Philadelphia, 862 F.3d 353 (3rd Cir. 2017).

Jones v. Opelika, 316 U.S. 584 (1942).

Nat'l Press Photographers Ass'n v. McCraw, 90 F.4th 770, 777 (5th Cir. 2024).

W. Virginia State Bd. of Educ. v. Barnette, 319 U.S. 624, 632 (1943).

Wooley v. Maynard, 430 U.S. 705, 714 (1977).

CHAPTER 2

The First Amendment Right to Record and Seeing as "Speech Creation"

Abstract If we ask what gives us a right against government interference in our *natural* vision—that is, our ability to turn our heads and look around us—the best answer is likely not the First Amendment. What protects us against government interference here is personal freedom and bodily autonomy. However, it is not a surprise that—when we move from natural seeing to seeing with technology—First Amendment freedom of speech has become the focus of American courts. This is because, as the Court held in a 1965 case, *Zemel v. Rusk*, we may have a right to see what is in front of us, but we don't have a right to ourselves everywhere we wish to gather information. To do that, we have to act as the audience for a speaker who is already there or has a right to be there. The right to record presents a partial answer to that problem—but an incomplete one.

Keywords Right to receive information · Freedom of speech · Freedom of thought · Right to record · Speech creation · Constitutional rights · Drone cameras · Supreme Court · American courts · Deepfake · Privacy · Fourth Amendment · Surveillance

When a Right to Perceive Becomes a Right to Perceive with Technology

Rights, including constitutional rights, are often understood as having the purpose of giving legal force to important interests: For courts recognizing "a right to record," that interest is in expressing ourselves with videos. Yet there is another interest that undergirds a right to see. We have a powerful interest in seeing the world for ourselves.[1] We have an interest, as we view the world, in being free of filters designed to tightly confine or distort our perceptions. That is true whether those who impose such filters are government officials or private actors. Or so I have argued before and will argue again more fully in this book. I have argued in the past that the intellectual liberty crucial to a free society "depends not only on our ability to communicate free of government monitors and censors," of the kind the First Amendment's free speech clause clearly forbids, "but also to observe the world free from government-imposed blinders."[2]

That we have a powerful *interest* in remaining free from restriction or manipulation of our perception does not mean, however, that we necessarily need a *right* (constitutional or otherwise) to ensure this freedom. A right against government (or other) interference in our perception is needed only if such interference is at all likely. When we view our immediate surroundings with our *natural* visual system, there is a case to be made that there isn't any such necessity: While government officials can certainly infringe our liberty interests, such infringements have rarely come in the form of forcibly blindfolding us or legally prohibiting us from looking at what is already visible to us. As John Humbach notes, "[v]ery rarely has the law imposed obligations on individuals to avert their eyes or stuff up their ears in order to avoid seeing or hearing other people or observing their belongings."[3] In cases on police searches, the United States Supreme Court has stressed that law enforcement officers cannot

---

[1] *See* Marc Jonathan Blitz, Freedom of 3D Thought, 1212–1213.

[2] Marc Jonathan Blitz, *The Right to Map (and Avoid Being Mapped): Reconceiving First Amendment Protection for Information-Gathering in the Age of Google Earth*, 14 Colum. Sci. & Tech. L. Rev. 115 (2013).

[3] John A. Humbach, *Privacy and the Right of Free Expression*, 11 First Amend. L. Rev. 16, 41 (2012).

reasonably be expected to "avert"[4] or "shield"[5] "their eyes" from what has been left open for them to observe in public spaces, where they have a right to be. The state generally doesn't demand *anyone else* avert or shield their eyes in that way. Moreover, where the government is able to interfere with our use of our eyes or visual brain processing, this would likely violate a right that is more familiar than "a right to see": It would violate our physical liberty and bodily autonomy.[6] So, one might argue, a distinctive right to see the world around us isn't needed. We find all the protection we need in other, more familiar rights.

The situation changes, however, when we see or otherwise sense the world *with technology*. And, in the current age, this is increasingly how we see many of the environments we need to explore and events we need to know about. In fact, in many cases, we can only see the world for ourselves by seeing it through a "window" of sorts. Since the rise of television in the mid-twentieth century and personal computers in the late twentieth century, we have increasingly viewed the world by watching images transmitted on screens. Our vision is increasingly mediated by technology. Such sensory mediation has become even more deeply ingrained in our day-to-day lives with the emergence of mobile phones, and the subsequent incorporation of computers—and computer screens—into these phones, along with internet connections that let us continuously receive audio, video, and images on web browsers and other applications.

Not only is such technology for receiving and viewing images ubiquitous. So too is technology for transmitting and capturing them. In urban landscapes—and increasingly other settings as well—powerful cameras are everywhere. They are being integrated not only into the cell phones we carry everywhere but also into other devices. We find them in the doors of houses that look out onto the streets, in massive government camera networks that vigilantly watch over cities, in drones that can hover over private as well as public properties, and in fleets of cars,

---

[4] California v. Greenwood, 486 U.S. 35, 41 (1988).

[5] See California v. Ciraolo, 476 U.S. 207, 213 (1986).

[6] See Terry v. Ohio, 392 U.S. 1, 8–9 (1968) (emphasizing the "the right of every individual to the possession and control of his own person"). Chapter 4 of this book explores in more depth how a right to physical integrity can provide a legal shield against disabling of our sensory perception.

planes, and other vehicles that Google and other companies with "mapping services" use to create massive photorealistic reproductions of public space that consumers can use to navigate or explore the world. As noted in Chapter 1, engineers have developed, and companies are now experimenting with, wearable cameras embedded in glasses or visors that might also double as "augmented reality" (AR) displays that can superimpose images or information over a person's field of view or enhance their visual perception.

When we see not only with our natural visual systems—but through channels and with tools generated by new technologies—our perception of the world undergoes at least three changes that make it more likely to be a target of government interference (or perhaps other outside interference). First, our vision might become more *vulnerable to manipulation or restriction*. When the sensory information that reaches us comes to us by way of channels far outside of our bodies—and far outside of our homes and other private spaces we control—it becomes far easier for government officials and other actors to hijack and block. If, for example, we rely on a video feed to see certain environments or events, those who wish to block our vision can sever the wires or intercept the transmissions that carry that feed to the screens we watch it with. They could even conceivably hijack the transmission and swap the images of the world we believe we are seeing with images they want us to see. As noted below, deep-fake videos allow them to disguise an animation of their own creation as camera footage of a real event. If our perception of reality is mediated by technology, it is at least possible that, instead of simply giving us greater control over how we see the world, it will allow officials or other private actors to exercise such control over what we see or how we see it.

Second, it becomes easier for government officials to interfere in our vision *without intruding into recognized spheres of personal autonomy*. At least when our vision is augmented with remote technologies—rather than an artificial eye, a brain-computer interface, or some other device implanted in, or connected to, our bodies—it is less tightly integrated with bodily systems and private spaces in which we have a strong and long-established claim against any outside interference. Government officials and other outside actors at the very least have a more plausible argument that their interference in what we see should be permissible when their target isn't our bodies and homes but rather technologies in shared public spaces or on others' property. Consider camera-equipped drones that fly over cities or farms, or other vehicles that can surveil stretches of public space, or wires and wireless channels that wind through public spaces. These are all technologies and places that have been regulated by governments in the U.S. and elsewhere aren't generally considered to be off-limits to such regulation on the ground that they are private or personal spaces of the kind from which government is presumptively excluded.

The third reason is related to the second: Not only does our use of technology to enable or augment our perception often move our vision to a realm where we have less autonomy—and thus are likely to suffer less harm from government intervention. It moves it to a realm where government officials and perhaps also members of the public may argue that there is *more need for such intervention to counter potential harm*. When our view is aimed not at our own homes—and we are not simply star gazing through telescopes or bird watching with binoculars but rather viewing the actions of others in public spaces, or perhaps at what they do in backyards or balconies from a vantage point in public space—then it is more plausible for officials to claim that we may sometimes see something we *shouldn't* have a right to see. This is particularly true if cameras that allow us to view such activities are supplemented with other technologies. A thermal imaging device might let someone see through the walls of an enclosure or the outer surface of a handbag, backpack, or suitcase. A zoom lens or other magnification technology might let a viewer discern small details—letting one person see what another person is reading or a text message they are sending. A parabolic microphone might let them listen to private conversations that spouses have beside an open window or while they speak quietly at a restaurant. When such a video camera or microphone is built into a drone that flies over or near private spaces, it may allow for even more significant invasions of privacy—especially if it is part of a fleet of drones equipped with cameras or microphones rather than a single drone operating by itself.

The possibility of such high-tech recording and observation has led numerous scholars—including the author of this book—to argue that the government's use of such surveillance technology cannot be left free from the constitutional protections designed to restrict "unreasonable searches."[7] In the United States Constitution, as noted earlier, the

---

[7] *See, e.g.*, Christopher Slobogin, *Public Privacy: Camera Surveillance of Public Places and the Right to Anonymity*, 72 Miss. L.J. 213, 236 n.106 (2002); Christopher Slobogin, *Privacy at Risk: The Government Surveillance and the Fourth Amendment*; Marc Jonathan

Fourth Amendment specifically protects "the people" against "unreasonable searches" of their "persons, houses, papers, and effects."[8] In recent decades, the Supreme Court made clear that government violates this provision when it uses technology to eavesdrop on a conversation in a private space,[9] to see through walls,[10] or to make or access massive personal archives showing everywhere we have been over a period of months (or longer).[11] But privacy threats can also arise when such technologies are employed by private companies—or even other individuals. Not surprisingly, government regulators have worried about the privacy implications of the repeated massive image capture Google has engaged in to build Google Streetview.

In recent years, there is also another threat that has emerged and brought calls for government intervention in the technologies we employ to see the world. Unlike the threat to the privacy of those who might be the subjects of observation, this is a threat to the perceivers themselves. With the rise of "deepfake videos" and the development of increasingly realistic VR experiences, it is increasingly possible for technologies we rely on to learn about the world to instead deceive us about it.[12] We might think we are seeing an accurate camera-captured record of a distant or past event when, in fact, we are seeing a fiction fabricated by artificial intelligence technology. Congress and many states within the U.S. have thus

---

Blitz, *Video Surveillance and the Constitution of Public Space: Fitting the Fourth Amendment to a World that Tracks Image and Identity*, 82 Tex. L. Rev. 1349–1481 (2004); Andrew Guthrie Ferguson, *Persistent Surveillance*, 74 Ala. L. Rev. 1, 7 (2022); David Gray and Danielle Citron, *The Right to Quantitative Privacy*, 98 Minn. L. Rev. 62, 65 (2013).

[8] The text of the Fourth Amendment reads: "The right of the people to be secure in their persons, houses, papers, and effects, against unreasonable searches and seizures, shall not be violated, and no Warrants shall issue, but upon probable cause, supported by Oath or affirmation, and particularly describing the place to be searched, and the persons or things to be seized." U.S. Const. amend. IV.

[9] Berger v. New York, 388 U.S. 41 (1967).

[10] Kyllo v. United States, 533 U.S. 27 (2001).

[11] Carpenter v. United States, 585 U.S. 296 (2018).

[12] Bobby Chesney and Danielle Citron, *Deep Fakes: A Looming Challenge for Privacy, Democracy, and National Security*, 107 Cal. L. Rev. 1753, 1759 (2019).

contemplated—and some states have already enacted—laws that prohibit deepfakes or require disclosures alerting viewers that they are fake.[13]

For all of these reasons, even if government is highly unlikely to interfere with the seeing we do when we scan our immediate surroundings with our eyes, it *is* likely to regulate the seeing we do with technology. In fact, there are circumstances where we may even want it to—where our interests in remaining unseen or undeceived seem to require that. However, if seeing the world for ourselves remains a deeply significant interest of ours even when we do it with technology, then there is good reason to think it should be the subject of a right or constitutional liberty protection which ensures that the government does it for legitimate purposes and in legitimate ways.

This chapter and the remaining chapters of this book will therefore explore what such a right might look like in American constitutional law, what its constitutional foundations might be, and how it might operate. As I noted in the introduction, I will ultimately argue that such a right should be understood, in large part, as a component of our freedom of thought or what American courts have sometimes called our "freedom of mind." In fact, the pattern I discuss above—in which our natural vision needs little in the way of rights-based protection but our technologically enhanced seeing needs robust protection—is one that I and others have argued applies to freedom of thought: Private beliefs and reveries have historically needed little shielding against government punishment because, as one judge put it, "the most tyrannical government is powerless to control the inward workings of the mind."[14] When, by contrast, we seek to shape our consciousness with technologies—such as chemicals that enhance our cognition or computers that let us shape how it works—this supplement to our thinking makes our shaping of our minds more vulnerable to government regulation, and more likely to be viewed as a legitimate target of it. The same is true, I am arguing here, of our perception and that is no coincidence. Our use of perceptual processes is

---

[13] For a good overview on proposed and enacted state law on deepfakes see Daxton R. Stewart, and Jeremy Littau, The Right to Lie with *AI? First Amendment challenges for state efforts to curb false political speech using deepfakes and synthetic media*, 2024 Annual Conference of AEJMC https://papers.ssrn.com/sol3/papers.cfm?abstract_id=4830972. Among the statutes that require disclosure for election related deepfakes are Michigan HB 5141 (2023), Texas Election Code Sec. 255.004(d) (Tex. 2024), H.B. 986 (Georgia 2023–24), and Cal. Elections Code 2100.10(a).

[14] Jones v. Opelika, 316 U.S. 584, 618 (1942) (Murphy, J., dissenting).

part of the way we exercise our minds: What and how we see is inextricably bound with the way we gather information and form beliefs about the world, learn to navigate it, and imagine alternative ways of living in it (or alternative versions of it).[15]

How then should we define such a right? The introduction to this book already provided one possible starting point: American courts have *already* had a lot to say about our right to see with technology in cases where they treated such technology as an essential part of the way we share information with each other. Video recording allows us to preserve a record of what we see—and enhance our seeing by, for example, watching the same event repeatedly, pausing on a particular frame, and magnifying detail in it. But the visual record also allows us to vividly communicate what we see. In this respect, sharing videos is an exercise of our freedom of speech—and our right to see is a component of our right to act as an audience or what we see, or to scrutinize and edit our own recordings so they can later share with them with others. Or perhaps use a photo or video the same way we use a journal entry that we choose *never* to share: To inform ourselves about the event or scene that it depicts and privately reflect upon it.[16]

The First Amendment right to make camera recordings of what we see and share those recordings with others, then, could possibly serve as a template for how courts might define the contours of a broader right to see with technology (or perhaps a broader set of distinctive rights to see with different kinds of technologies). In the remainder of this chapter, I will therefore briefly explore some of thethe First Amendment case law and doctrine on information gathering and recording to test this proposal. And I will conclude that while such a First Amendment right to record is invaluable—and answers some of the challenges that confront a right to see with technology—it is far from a sufficient foundation for such a right. Finding such a grounding requires building it not solely upon free speech doctrine but upon another constitutional foundation—most notably, a broader constitutional protection for freedom of mind.

---

[15] Tyler Burge, *Perception: First Form of Mind* (Oxford University Press 2012). That perception is related to thinking doesn't mean that our use of perception should be equated with the latter. But to the extent certain internal mental activity depends on perception, both may fall within the scope of the same right.

[16] *See* Martin H. Redish, *Freedom of Thought as Freedom of Expression: Hate Crime Sentencing Enhancement and First Amendment Theory*, 11 Criminal Justice Ethics (1992).

Zemel v. Rusk, The First Amendment, and Our Lack of an "Unrestrained Right to Gather Information"

The First Amendment case law on video recording and other image capture is quite new. Courts have long held that motion pictures[17] and artistic photography[18] count as speech protected by the Constitution. It is only in the past two decades, however, that many of them have found that image capture and audiovisual recording more generally is protected. In recent years, as noted in Chapter 1, numerous American courts have found that individuals have a right to make camera recordings—at the very least of police encounters and other matters of public concern.[19] But I want to begin this chapter by exploring how the Supreme Court analyzed another, older technology that can augment our perception: Air travel and other means we can use to change *where* we see the world from.

If we want to perceive and learn about particular events and environments, after all, it is not sufficient that we be able to use our eyes and natural visual systems to scan our immediate surroundings. We also have to be able to *locate ourselves* so that our immediate surroundings contain the information we wish to learn. The latter claim, said the Supreme Court, cannot plausibly be rooted in the First Amendment: We may have a right to make, and later share with others, observations where we already have a right to be. But that doesn't mean we can claim a right to be anywhere that we wish to make observations.

The Court made this point in a 1965 decision. In that case, Louis Zemel brought a constitutional challenge to a decision by the U.S. State Department effectively denying him the chance to visit Cuba. To go there in 1962, when he wished to make the trip, he needed not only a passport but also the State Department's authorization that he could use his

---

[17] See Burstyn v. Wilson, 343 U.S. 495 (1952) (finding that motion pictures are speech protected by the First Amendment).

[18] See Kaplan v. California, 413 U.S. 115, 119–20 (1973) ("[P]ictures [and] films ... have First Amendment protection").

[19] *See, e.g.*, Fordyce v. City of Seattle, 55 F.3d 436, 439 (9th Cir. 1995) (speaking of a "right to film matters of public interest"); Askins v. U.S. Dep't of Homeland Sec., 899 F.3d 1035, 1043 (9th Cir. 2018) (speaking of the same right); Irizarry v. Yehia, 38 F.4th 1282, 1289 (10th Cir. 2022) ("[f]ilming the police and other public officials as they perform their official duties acts as "a watchdog of government activity"... furthers debate on matters of public concern").

passport for that purpose, something the State Department refused to give him. Relations between the U.S. and Cuba's communist government were hostile and even more tense than they are now. Indeed, as the Court observed in its decision, the Cuban missile crisis of October 1962 occurred soon after Zemel's request was refused and "preceded the filing of appellant's complaint by less than two months." So the State Department argued it had national security reasons to deny most Americans authorization to travel to Cuba.[20]

Zemel's constitutional argument essentially made two claims. First, he cited prior cases that had recognized that individuals have a right to travel within the United States (as part of the protection of "liberty" provided by the Fifth Amendment)—and argued that the right also protected international travel. The Supreme Court took that argument seriously: It acknowledged, in its decision, that "[t]he right to travel within the United States is of course [] constitutionally protected."[21] It also suggested that such a right might apply, at least in some circumstances, to international travel as well. Still, it said, this did not mean that this right was violated by the restriction on travel to Cuba. Travel can be limited, the Court stressed, to protect the "safety and welfare" of the United States or some part of it. That is why, it said, the right to travel does not prevent the government from placing off-limits "areas ravaged by flood, fire or pestilence."[22] Nor does it prevent the government from restricting travel where doing so is justified by "the weightiest considerations of national security," such as the considerations the State Department relied upon in this case.[23]

Zemel's second claim invoked the First Amendment. The right to freedom of speech, Zemel claimed, not only protected his right to learn about the world from reading what others have written or hearing what they have said about a particular location or about current events. It *also*, his legal brief argued, gave him and other individuals a right "to travel abroad so that they might *acquaint themselves at first hand* with the effects abroad of our Government's policies, foreign and domestic, and with

---

[20] Zemel v. Rusk, 381 U.S. 1, 15 (1965).
[21] *Id.*
[22] *Id.*
[23] *Id.*

conditions abroad which might affect such policies."[24] This argument—that the First Amendment's shield extends to a person's *direct contact* with the world—had also been defended, as Zemel noted in his brief, by others in the legal community. He cited an article in Atlantic Monthly written in 1952 by a federal judge, Charles E. Wyzanski, Jr., arguing that the right to travel should be classified as one of "the manifold freedoms of expression" alongside the "right to speak, to write, to use the mails, to publish, to assemble, to petition" because all of these liberties allow for "the stretching of the mind to accommodate the growing spirit."[25] He noted that a 1958 Report by a committee of the New York City Bar had also said the freedom to travel serves crucial First Amendment interests: "[F]reedom to travel," it had said, "is a right closely related to First Amendment interests" and for this reason should only be denied in extraordinary circumstances.[26]

The problem with this argument, said the Court, is that it leads to an absurd conclusion. It was certainly true that the "Secretary's refusal to validate passports for Cuba renders less than wholly free the flow of information concerning that country."[27] But virtually *any* legal restriction could be attacked on essentially the same grounds. *Every* barrier the government places on our freedom of action, after all, prevents us from seeing or otherwise perceiving certain things—some of which may be valuable for citizens to see. Entry into The White House may enable a person to observe and learn about the President's conversations with other officials.[28] But that can't mean every citizen has a right to enter it at any time. "There are few restrictions on action," the Court stressed, "which could not be clothed by ingenious argument in the garb of decreased data flow."[29] That can't mean that every restriction on action

---

[24] *Id.* at 16 *(emphasis added) (emphasis added) (quoting Brief of the Appellant (Louis Zemel)*, Zemel v. Rusk, *1964 WL 81305, at *51 (U.S., 2004)*.

[25] Brief of the Appellant (Louis Zemel), Zemel v. Rusk, 1964 WL 81305, at *51 (U.S., 2004) (quoting Charles E. Wyzanski, Jr., *Freedom of Travel*, The Atlantic, Oct. 1952, at https://www.theatlantic.com/magazine/archive/1952/10/freedom-to-travel/642063/).

[26] *Id.*; Freedom to Travel (Report of Special Committee to Study Passport Procedures, Assn. of the Bar of the City of New York, 1958).

[27] Zemel v. Rusk, 381 U.S. 1, 16 (1965).

[28] *Id.*

[29] *Id.* at 16–17.

is subject to a plausible First Amendment challenge. The First Amendment gives us a right to freedom of speech, not a right to take *any* action we might wish to take that will generate particular perceptions we otherwise wouldn't have. For this reason, said the Court, "[t]he right to speak and publish does not carry with it the unrestrained right to gather information."[30]

The Court's focus here, of course, was on analyzing the First Amendment's free speech clause and the argument that it includes a right to gather information. However, one might also recast the lesson of this case as an answer to the question I raised earlier in the chapter: When does our unquestionable right to observe our immediate environment with our natural vision entail a right to change our situation so that we can see further or more deeply? One way of changing our situation, of course, is by transforming our vision—and the processing of that vision—with technology. Bionic eyes of the future and high-powered magnification on cameras can let us see things we could never see with our unaided vision.[31] But there is also a simpler, "low-tech" way of changing our situation to allow us to see what we otherwise couldn't see—and that is by changing our location. Movement, as noted earlier, changes what is *in* our immediate environment and thus changes what we can see when we look out at it. Indeed, our use of natural vision—like that of other animals—generally happens in conjunction with movements that enable us to focus our attention where we believe it needs to be focused. When we need to see something behind us, we turn our heads or reposition our entire bodies. Perception doesn't just entail soaking up sensory information with the eyes but using our motor capacity to move our eyes—and sometimes other parts of our body—to change the vantage point from which we see and sense.[32]

---

[30] *Id. at 17.*

[31] *See* Marc Jonathan Blitz and Woodrow Barfield, *Memory Enhancement and Brain-Computer Interface Devices: Technological Possibilities and Constitutional Challenges*, in ed. Veljko Dubljevic and Allen Coin, *Policy, Identity, and Neurotechnology* (Springer 2023).

[32] *See* Christoph Koch, *The Quest for Consciousness: A Neurobiological Approach* (Roberts & Co; 2004), p. 63 (noting that "[y]ou move your eyes all the time. You read by skipping with small saccades ['rapid movements of both eyes yoked together'] across the page. You look at a face by constantly glancing at its eyes, mouth, ears, and so on").

In a system of social relations where we have obligations to others—where we have to respect their property and privacy rights as well as certain other constraints, rooted in the public interest, on where we can be and when—we can't plausibly be able to insist that we be able to perceive the world from any location we wish to place ourselves. In fact, courts have even considered the possibility that even if we have a right to be in a place for *certain purposes* that may not necessarily give us a right to make recordings or otherwise conduct surveillance there: The fact that we have a right to walk in a public street or be a passenger in a helicopter in a certain area of the sky doesn't necessarily mean we have a right to use a camera in those places to magnify and capture images of small details of the activities people are engaged near the windows of a home or private business.[33]

Not surprisingly, many of the legal disputes over surveillance technology are, in part, battles over whether state officials or private individuals can make observations from certain vantage points. This is a question courts have asked about police surveillance in Fourth Amendment cases: Does the U.S. Constitution's bar on "unreasonable searches" in the Fourth Amendment raise any legal barriers against them using an airplane to scrutinize (for drugs) a backyard that would otherwise be hidden behind a fence?[34] Does it bar them from using a helicopter to position themselves above the roof of a greenhouse so that they can see through a crack in the roof and detect any drugs grown there?[35] Or from recording the conversations that occur in a house when the government investigator or informant hides both their affiliation with the government and the fact that they are recording the private statements by the target of the investigation?[36]

It is also a question that is increasingly raised in First Amendment cases on the right to create video recordings or audiovisual records. Multiple federal cases on this right have generally recognized individuals are protected by the First Amendment when they record police

---

[33] *See, e.g.*, Project Veritas v. Schmidt, 125 F.4th 929, 955 (9th Cir. 2025) noting that states have a "significant interest in protecting private conversations includes private conversations that occur in public or semi-public locations. There is little doubt that private talk in public places is common (citation omitted)."

[34] *See* California v. Ciraolo, 476 U.S. 207 (1986).

[35] *See* Florida v. Riley, 488 U.S. 445, 452 (1989).

[36] *See* United States v. White, 401 U.S. 745, 751–53 (1971).

performing public duties in public places.[37] But other cases have asked if individuals could surreptitiously record inside of a home when the homeowner permits them entry—or whether activists and journalists could expose wrongdoing by surreptitiously recording how dairy companies treat animals,[38] for example, or recording conversations they have with interest group member while posing a supporter.[39] Or use drones to gather information about how private companies' operations may be threatening the environment.[40]

One lesson to draw from the Court's decision in *Zemel v. Rusk*, then, is that questions of this sort about *where* we can perceive from, and whether there are limits on our right to record or observe even in the places where we have a right to be, cannot be answered simply with the response that we *always* have a First Amendment right to observe *anything* we might learn from. In a world where our observation and information gathering have to be reconciled with others' right to property and privacy, a right to see and learn about our environment with technology necessarily has to be more nuanced.

I therefore want to examine, in the remainder of this chapter, some proposals for how we might find such nuance in First Amendment doctrine on speech and receipt of information. As I have noted before, I will ultimately argue that we have to go beyond such doctrine in defining what rights we have to see with technology (and in some cases, remain unseen and undeceived by it). But I will start here with the First Amendment law on enhanced seeing that I began with the introduction: First Amendment law on the right to record. As I have just done with *Zemel v. Rusk*, I'll start by reviewing some of this law in the terms that courts use in applying and defining it and I will then translate this shielding for the right to record in different terms that relate it more clearly to the question I am focusing on here—namely, when do we have a right to see not only with our natural vision (in our home or other place where our

---

[37] *See*, for example, ACLU of Ill. v. Alvarez, 679 F.3d 583, 595 (7th Cir. 2012); Irizarry v. Yehia, 38 F.4th 1282, 1289 (10th Cir. 2022), Fields v. City of Philadelphia, 862 F.3d 353, 360 (3d Cir. 2017), Turner v. Lieutenant Driver, 848 F.3d 678, 690 (5th Cir. 2017), and Glik v. Cunniffe, 655 F.3d 78, 82 (1st Cir. 2011).

[38] Animal Legal Defense Fund v. Wasden, 878 F.3d 1184, 1191 (9th Cir. 2018).

[39] Planned Parenthood Fed'n of Am., Inc. v. Newman, 51 F.4th 1125, 1131 (9th Cir. 2022).

[40] National Press Photographers Ass'n v. McCraw, 90 F.4th 770 (5th Cir. 2024).

personal autonomy is at its strongest) but also with technology, and from vantage points where our freedom is in certain respects more limited. As I will explain, the right to record provides an answer of sorts to the difficulty that *Zemel* leaves us with: Under *Zemel*, we don't have a right to locate ourselves so as to learn anything we wish to learn from a distant environment. In a world where video recording is both widespread and protected by the First Amendment, however, we *do* have a good chance of viewing such a remote location as long as there is *someone* there who *does* have a right to be there and can create a record of what they see and hear—to let us see and hear it too.

THE RISE OF A RIGHT TO RECORD

The Supreme Court said in 1943[41] and repeated many times after that,[42] that the First Amendment includes a right not only to speak but also to receive information and ideas. However, it described this right in 1976 as a right to receive information only "from a willing speaker."[43] Requiring a link to speech helps answer the *Zemel* Court's concern: The government does not violate the First Amendment in every case where it causes "decreased data flow."[44] Virtually every law does that—and a system where every law carried great risk of violating the First Amendment would be one where representative institutions find themselves paralyzed. Every law they enacted would be subject to a First Amendment challenge. Rather, it is *only* laws that limit *certain channels* of information that the First Amendment safeguards to assure minimal conditions for intellectual liberty and democratic discourse—and these channels generally involve communication or other expression. The government may be able to stop us from visiting a certain country or entering "an area ravaged by flood, fire or pestilence."[45] It *cannot* constitutionally stop us from speaking or writing about it or listening to anyone else talk about it. It

---

[41] Martin v. Struthers, 319 U.S. 141, 143 (1943).

[42] *See, e.g.,* Lamont v. Postmaster Gen., 381 US 301,308 (1965); Kleindienst v. Mandel, 408 U.S. 753, 762 (1972); Denver Area Educational Telecommunication Consortium Inc. v. FCC, 518 U.S. 727 (1996).

[43] Va. Pharmacy Bd. v. Va. Consumer Council, 425 U.S. 748, 756 (1976).

[44] Zemel v. Rusk, 381 U.S. 1, 15 (1965).

[45] *Id.*

can limit (possibly harmful) physical encounters with those places. But it may not, under the First Amendment, stop us from creating and sharing, verbal descriptions or visual depictions of those places if we are already there and have a right to be there. Nor can it stop us from receiving such descriptions or images from others (who are there) if we cannot be there ourselves.

If *Zemel's* firm rejection of an "unrestrained right to gather information" seems to raise a barrier to our use of enhanced perception to learn about the world, what I have just said seems to lower it in a meaningful way: A would-be traveler to Cuba cannot claim a First Amendment right to *physically* go there. But he can claim such a right to *see and hear* the sights and sounds of that place if these perceptions can be delivered to him as part of a *communication* from a person who *is* there. Even if he can't have a physical presence in a restricted area, then, he might have a kind of *virtual* presence made possible in a partnership of sorts with a speaker who is already there. Indeed, in the age of social media, such an on-the-ground partner may be much easier to find. They might be a complete stranger, someone who posts a video of a certain destination that can be viewed by millions of people who are on the same social media platform.

Current case law on the "right to record" makes this possible. In the United States, numerous federal courts have now ruled that the First Amendment gives individuals a right to create recordings of their surroundings—at least when those recordings take place in public spaces or other spaces where they have a right to be and where they depict "matters of public concern."[46] One of the most influential and well-known of these rulings was a 2012 ruling by the Seventh Circuit Court of Appeals in a case called *American Civil Liberties of Illinois v. Alvarez*.[47] The case concerned an Illinois law that punished "eavesdropping." Illinois had made it a crime for anyone to make an audio recording of a conversation between two or more persons without the consent of every party

---

[46] Alan K. Chen and Justin Marceau, *Truth and Transparency: Undercover Investigations in the Twenty-First Century* 168 (Cambridge University Press 2023) (noting that multiple courts have found "a First Amendment right to record matters of public concern even on private property"). *See also* Fordyce v. City of Seattle, 55 F.3d 436, 439 (speaking of a First Amendment right to film matters of public interest"); Askins v. U.S. Dep't of Homeland Sec., 899 F.3d 1035, 1043 (9th Cir. 2018) (similarly speaking of a right to record matters of public interest).

[47] ACLU of Ill. v. Alvarez, 679 F.3d 583 (7th Cir. 2012).

being recorded. In fact, the law made clear, it was illegal even if *no* party to the conversation regarded it as "private."[48]

The civil rights group, the American Civil Liberties Union (ACLU), realized that this ban on recording of conversations stood in the way of a program it was initiating to "promote police accountability." The ACLU planned to make audiovisual recordings of the interactions that law enforcement officers had with members of the public—both as they patrolled or responded to calls for law enforcement in public streets and in political events, such as rallies or protests, organized by the ACLU. The recordings were not, the ACLU made clear, designed to invade the privacy of officers or of those they interacted with: They would include video and audio of police only when they were performing "their public duties" in "public places" and speaking "at a volume audible to the unassisted human ear." Still, none of these factors exempted the recordings from the Illinois ban.

So the ACLU sued. It asked a federal court to declare the Illinois provision unconstitutional under the First Amendment and to stop the law from being enforced against its program for recording police. The first federal court to hear the case rejected this claim. As noted above, the Supreme Court has said that we don't have an "unrestrained right to gather information" wherever we might find it. Rather, we only have a right to receive information from a "willing speaker." Where police officers don't wish their conversation to be recorded, said this federal court, they are not "willing speakers"—even when they are performing their jobs on public streets. To be sure, the receipt and capture of auditory information here had a closer connection to speech than Zemel's vaguely defined plan to learn about Cuba. The ACLU's specific aim was not only to video record what police officers did but also to record what they said. The law that stood in its way specifically criminalized recording of "a conversation." Still, the framework the court used in ruling against the ACLU was the one I have drawn upon discussing *Zemel* and the limits on the right to receive information: Someone audio recording others isn't saying anything themselves. Nor are they the intended audience of the people they are recording who, if they haven't given their consent to be recorded, may not be speaking for any audience. One does not engage in speech when one steals—from a purse or backpack—a letter addressed

---

[48] *Id. at 606.*

to someone else or a draft of a story the writer has never shared. That the letter or story is unquestionably speech doesn't mean any act one takes to access and read that speech is protected by the First Amendment. Although the court didn't analogize audio recording to that kind of action, it seemed to view it in the same way: Specifically, it found that the ACLU lacked "standing" to bring the challenge before a court because, in the absence of a "willing speaker" in its recording, it could point to no First Amendment injury.[49] "Police officers and civilians may be willing speakers with one another," it said, "but the ACLU does not allege this willingness of the speakers *extends to the ACLU*, an independent third party audio recording conversations without the consent of the participants."[50]

The court to which the ACLU appealed, the Seventh Circuit Court of Appeals, viewed the claim through a different lens. As the Seventh Circuit put it, this case shouldn't be analyzed as a "right to receive" case where a listener claims the right to receive the communications of a speaker. Rather, it implicated "a different set of First Amendment principles."[51] This is because video and audio recordings aren't simply a method of receiving information. They are also a means of *creating* speech. "The act of making an audio or audiovisual recording," explained the Seventh Circuit, has to be protected because otherwise the unquestionably protected "right to publish or broadcast an audio or audiovisual recording would be insecure, or largely ineffective."[52] In other words, speaking or other expression is often a process that stretches over a period of time—not a single, instantaneous action.[53] Observers trying to identify what parts of a person's or organization's conduct count as "speech" under the First Amendment often focus on the specific act that a person takes to share ideas or information they possess with an audience—for example, the act of posting a video or verbal message on social media. However, this is only one component of the entire speech process. An

---

[49] Am. C.L. Union of Illinois v. Alvarez, No. CIV.A. 10 C 5235, 2011 WL 66030, at *4 (N.D. Ill. Jan. 10, 2011), rev'd, 679 F.3d 583 (7th Cir. 2012).

[50] *Id.*

[51] ACLU of Ill. v. Alvarez, 679 F.3d 583, 592 (7th Cir. 2012).

[52] *Id. at 595–596.*

[53] *See* Ashutosh Bhagwat, *Producing Speech*, 56 Wm. & Mary L. Rev. 1029, 1029 (2015).

earlier stage of that process necessarily entails the creation of the video or message that the speaker intends to share. It was that kind of creation of informative videos that the ACLU wished to engage in—prior to publishing the video on a website in order to help further police accountability. Contrary to the first court's analysis then, the ACLU wasn't asserting a right to act as an *audience* of *someone else's* speech under the First Amendment. Had the ACLU been nothing more than an audience for police officers' words, its First Amendment right to record would have existed only if police were *willing* to share their words with the ACLU. Its right to record was part of the right that the ACLU had as a *speaker* to create and share its *own* speech about the police conduct it observed. That kind of speech creation, the Seventh Circuit made clear, is generally something the First Amendment protects even if those the speech is about haven't given their permission for it to occur. Newspaper reporters, for example, don't need the permission of government officials to write and publish an article conveying information about (and perhaps criticism of) what these officials have done. The First Amendment ensures that journalists can publish stories about government conduct *without* officials reviewing and censoring any newspaper stories they find unflattering or that contain information they'd prefer to hide from the public. Nor they can effectuate such censorship by banning reporters from using a pen and paper to write down their observations about government officials to provide the basis for such a story. The ACLU's audio and audiovisual records of police, said the Seventh Circuit, were protected for the same reason. It was a part of the way the ACLU planned to create speech about a matter of public significance.

This kind of analysis thus answers the challenge raised by *Zemel v. Rusk*: Photographic image capture, video recording, and audio recording of things other than speech—or of unwilling speakers—don't depend for their First Amendment status on an "unrestrained right to gather information" of the kind the Court squarely rejected in *Zemel*. Those who create such visual or auditory records aren't merely information-gatherers. They're *speakers*—or more specifically, individuals who have begun to create speech that they can later share.

Numerous other courts have embraced the same logic for protecting audio recording, video recording, and image capture of all kinds. In *Irizarry v. Yehia*, for example, the Tenth Circuit Court of Appeals stressed that "videorecording is 'unambiguously' speech-creation, not mere conduct."[54] In *Turner v. Lieutenant Driver*, the Fifth Circuit likewise said that since "the Supreme Court has long recognized that the First Amendment protects film," it follows that "the First Amendment protects the act of making film."[55]

This logic for recognizing a right to record is certainly compelling. It is also akin to other judges' emphasis on the need for the First Amendment to protect a speech process—and certain conditions that make it possible—rather than simply the act of expressing words (or other communicative content) at the moment they are shared. The need to view free speech protection this way was, for example, a central part of Justice Douglas's opinion in the seminal Supreme Court case of Griswold v. Connecticut in 1965. That case established that individuals had a constitutional right to use contraceptives.[56] In doing so, it made the general point that constitutional rights can't be narrowly interpreted to cover only the most specific variant of the conduct they protect. Rights often have to be understood more broadly if they are to have any force. The First Amendment's protection for speech, wrote Douglas, provides an example. Interpreted narrowly, its protection for speech may seem to extend only to what people do when they "utter or print" words. But such narrow protection for only the acts of uttering and printing would leave our communication open to all kinds of other government measures that could effectively silence us: Government could attack thinking, reading, and education before they give form to our ideas, or prevent us from generating the content we would later utter or print.[57] It could stop libraries, bookstores, or others from distributing printed ideas after they are published. A properly conceived right to free speech thus has to protect the *whole process* by which speakers generate speech and then deliver it to listeners or readers. As Justice Douglas put this point: Without "peripheral rights" to distribute and receive information, to think, and to read, "the specific rights" to "utter and print" "would be less secure."[58]

---

[54] Irizarry v. Yehia, 38 F.4th 1282, 1289 (10th Cir. 2022).

[55] Turner v. Lieutenant Driver, 848 F.3d 678, 689 (5th Cir. 2017).

[56] Griswold v. Connecticut, 381 U.S. 479 (1965).

[57] *Id.* at 482–483.

[58] *Id.*

Other First Amendment cases have extended protection to other essential conditions of speaking. For example, a speaker cannot speak effectively unless they have someplace to speak from. Especially in the days before the internet, they often needed to reach members of the public in some physical location where they could encounter them—such as streets or parks. If the government could use its control over such public property to adopt rules barring speakers from sharing controversial views there, it could effectively silence them by, in a sense, pulling the ground out from underneath them. This problem isn't solved by a recommendation that speakers find some other place (like their own property) to use as a platform for speaking to the public. Before the rise of social media and internet communication, they could not easily reach the public from the private space inside their own home. They needed access to *public* spaces owned and operated by the government. So courts developed a specific First Amendment doctrine—now generally described as "public forum doctrine"—to ensure that access.[59] Similarly, individuals can't engage in certain types of communication without spending money or other resources to fund a messaging effort. So the Court developed First Amendment doctrine to ensure that government couldn't silence such communication efforts by barring the spending or other use of resources necessary to make them happen.[60]

Courts have likewise stressed that protection of speech may entail protecting the tools that allow individuals to create it. As the Ninth Circuit Court of Appeals has noted, the First Amendment couldn't protect musical and artistic expression if it only protected such expression at the moment it was shared in a musical performance or the display of a painting. Such narrow protection for art would let the government undermine it by, for example, making it illegal to produce or sell the instruments musicians need to make music or the brushes and canvases artists need to paint.[61] Video and audio recordings are protected for the

---

[59] It set out this doctrine in Perry Educ. Ass'n v. Perry Local Educators' Ass'n, 460 U.S. 37, 45 (1983). There it said that in streets and parks and other places "which by long tradition or by government fiat have been devoted to assembly and debate, the rights of the State to limit expressive activity are sharply circumscribed."

[60] *See* Buckley v. Valeo, 424 U.S. 1, 19 (1976). There the Supreme Court notes that "virtually every means of communicating ideas in today's mass society requires the expenditure of money."

[61] Anderson v. City of Hermosa Beach, 621 F.3d 1051, 1061–62 (9th Cir.2010).

same reason—and so might be the creation and distribution of at least some of the cameras and microphones that make them possible.

This then provides a First Amendment account of why and how the US Constitution protects at least one kind of right to see with technology—namely, a right to capture images and sound for later scrutiny or sharing. This right to see with technology is one component of a larger right to *speak* with technology. And since the First Amendment unquestionably protects the latter right—it allows us to express ourselves to audiences on social media and the web—it simultaneously protects the former.

Rights to Record as a Right to Enhanced Seeing (and Thinking)

We can also think of the First Amendment right to record described in these cases in slightly different terms. It isn't only a right to *speech creation*. It is also a means of enhancing and extending our right to perceive the world even *outside* the realm where our individual autonomy is at its strongest—the environment inside of our own home or other property.

Thanks to modern technologies of communication, we can exercise perception—in a manner protected against government intervention—not only by looking out at what we can see around us but also at the view that others give us of the environments that they interact with, in video or audio recordings they share with us or perhaps livestream they send to us. We have a right not only to observe the world directly but to act as an audience for others' communications about it—and, in the twenty-first century, these communications can not only take the form of words or symbols but also come in the form of records of light and sound they share with us after capturing them in cameras or other recording technologies.

This aspect of twenty-first-century discourse allows us to use it not only to hear what others think and claim but also to *see what they see*—to transport ourselves, at least in a limited way, to where they are and see and hear the world from their physical vantage point. In doing so, we not only see and sense from a physical perspective different from our own. We can, as one court has pointed out, *enhance* our seeing—we can make it more effective. "[T]o record," the Third Circuit Court of Appeals has said, "is to see and hear more accurately."[62] A recording "corroborates or lays aside subjective impressions for objective facts."[63] At least one kind of technologically mediated seeing, then, is constitutionally shielded along with our own immediate perception of our surrounding environment: The seeing we do with the technological mediation used by *other people to convey information to us* is covered by the shield given to First Amendment expression.

As I noted earlier, this provides an answer of sorts to the problem that confronted Louis Zemel and that confronts many other individuals who want to *see for themselves* certain environments they are not allowed to physically be in. The answer lies in the fact that, at least in the twenty-first century, seeing isn't merely a solitary activity that each individual can engage in solely from their own physical vantage point. It is a social practice that we engage in as we share information with others and receive information from them in return. In the past, such sharing could only be accomplished with verbal testimony about, or perhaps artistic illustrations drawn by, an onsite observer. In an age of pervasive video recording and sharing, by contrast, that observer can allow us to essentially see and hear a portion of what she sees. In a future where VR technology is more widespread and advanced, she may even be able to give us "telepresence" there: To let us immerse ourselves in a convincing illusion of being transported there even though we are physically far away.[64] In such a communicative endeavor, our right to see the environment is not limited to a space (such as the home) where our own claim to autonomy is at its height. It can also extend to places where *others'* exercise of *their* autonomy, and *their* right to see, can capture additional sensory information for us. Recording technology and the First Amendment rights that shield it thus allow a right to see to be enhanced and extended even in a world where the government has—and has to have—power to limit our physical activity in some respects, and the surveillance we conduct in the course of it, outside of the home. Where we reach a limit of our own

---

[62] Fields v. City of Philadelphia, 862 F.3d 353, 359 (3d Cir. 2017).

[63] *Id.*

[64] *See* Sin Duc Hoang, et al., *Harnessing the Power of Virtual Reality: Enhancing Telepresence and Inspiring Sustainable Travel Intentions in the Tourism Industry*, 75 Technology in Society, Nov. 2023 (exploring how "investigates how virtual reality (VR) can enhance telepresence").

autonomy, we can still share in—through communication of images and videos—the perceptions available in *someone else's* realm of autonomy.

To be sure, many questions follow from this idea that we have a right to see what others have a First Amendment right to record and share. The most challenging puzzles involve defining what it is that others have such a right to record. That may sometimes be straightforward. Others have a right to record and share recordings of their own personal activities in their home or other places they have a right to be. It seems clear that they have a right, for example, to post social media videos demonstrating how to cook a particular recipe, make a repair to their home, or perform a kind of physical exercise, or showing themselves playing a video game or musical instrument. As noted above, courts have also found that they have a First Amendment right to record and share "matters of public concern" such as police activity in public spaces. It is less clear whether and when they ever have a right to make and then share recordings on others' private property—such as a business that they sneak into or, even if they are an employee with permission to physically be on the business's premises, when there are employment rules barring any such recording. As noted in Chapter 5, some scholars have argued, and some courts have found, that even recording on private property without the owner's permission should receive First Amendment protection as long as the recording reveals "matters of public concern." In any event, whatever line courts draw to reconcile the First Amendment interests of speakers and audiences and the interests of those whose private activities or spaces might be recorded and revealed without permission, the right to record provides a solution for balancing these interests—and doing so in a way that could not be achieved by an "unrestrained right to gather information."

Why then isn't this a sufficient answer to the question of when and how we can extend our powers of perception? Thanks to the First Amendment, we can do so where others are willing and able to make their perceptions our perceptions—with the aid of recording technology. The problem is that it is too narrow. Certain forms of technologically enhanced seeing do not involve communication. Communicating is not their point. Rather, they are designed to let us see for ourselves *without* someone else's choices mediating what see. The extension of our seeing is instead accomplished *solely* by mechanical means of some kind—and largely shuts out anyone else's active decision-making. Think, for example, of what occurs when a person places a camera on a drone and then either receives a video feed from the drone as they pilot it through the air, or watches a recording retrieved from the drone after it returns. They are not, in this case, viewing a video feed someone else is shaping or a recording someone else has made. They are watching video footage that comes only from a camera they have set up and operated themselves. In doing so, they extend their seeing without anyone else being involved. Is *this* form of extended seeing constitutionally protected even though it does not entail any exchange of information between speakers and listeners?

In fact, even when someone else has set up or is operating the drone, it is not clear that this makes us an audience for their speech in the same sense we act as such an audience when we listen to something they say, read a report they have written, or view an artwork they have created. Imagine that I transmit photograph images or a video to a viewer—someone, in other words, who is acting as my audience—from a camera-laden drone I have programmed to circle above a particular area of interest to that audience. If communication is, to use Justice Robert Jackson's memorable phrase, a "short cut from mind to mind" is that what the drone camera is creating? That would certainly be the case if I not only operated the drone carrying the camera but also edited or otherwise exercised some authorship over the images being transmitted. But is it the case when I've simply set up the technological apparatus that transmits the images to the viewer? It seems not to be. First, it's possible for me to arrange for this kind of image capture and transmission without ever viewing the images captured by the drone myself. Even if I view them at the same time as my audience receives them, the shortcut is better described as one that is established between a machine and both of our minds: Even if I have programmed the drone's operation, I seem more like another audience for the images it sends than a speaker who authors them. Moreover, the experience the other viewer has is identical to the experience they would have had if they had programmed the drone's movements and image capture themselves. It is odd then to describe this kind of image transmission as being perfectly analogous to communication between a speaker and an audience. The nature of the perception that occurs here is more like the perception a person has of stimuli that come to them from the surrounding physical or built environment, perhaps with the mediation of a camera and computer, than the perception they have of the verbal or other expression made by a human speaker. Consequently, even though what I have described counts as the sharing of live-streamed or recorded images that involves two people—one who has initiated the process and is perhaps directing the mechanics of it—the underlying interest might be better described as the same kind of interest that would underlie perception that a single person makes of their environment.

As I have already noted in the introduction, and will explain more fully later, even when a single person makes a recording *without* any plan to share it, courts now generally treat such an act as an act of "speech creation," just as they do so when someone writes a poem or essay they never publish or even show anyone else.[65] But that does not fully answer the need to analyze this kind of extended seeing. It doesn't explain, for example, how one should treat our use of video feeds to see at a distance, and perhaps have VR-enabled telepresence there, when these feeds leave no record of the experience except in the mind of the perceiver. Nor does it address when and where people should be able to enhance their vision by perceiving radiation or other stimuli they normally cannot see.

Moreover, when courts treat the value in a recording as inhering only in our potential to share it with others, they miss an important facet of it. A video or audio recording can certainly be a valuable component of our communications with others. But it might also be valuable when it lets us do *without* such communications—when it *frees* us from them. The perception that machines enable can be designed to happen without anyone else's intervention and to remain unshared with others (until we allow for that sharing to happen). In other words, this form of seeing with technology is not only enhanced but also solitary. It allows us to see for ourselves.

To be sure, there are questions that confront any claim that we should have a right to this kind of *non-communicative* enhancement of our perception. One is that it is harder to root in any constitutional text or tradition than is the enhanced seeing one does as part of a communicative process. It is possible, as I will explain in Chapter 4, that courts could treat such seeing as an extension of bodily autonomy (protected as part of the personal liberty shielded under the Fifth and Fourteenth Amendment and the Fourth Amendment's protection against unreasonable searches and seizures). Yet although such an extension intuitively makes sense for visual prosthetics or other technology that is embedded in one's body,

---

[65] Fields v. City of Philadelphia, 862 F.3d 353, 360 (3d Cir. 2017).

or physically linked to it, it is harder to justify applying it to technology that is miles away from one's person—such as a drone one is piloting or a camera in a distant country collecting light and sound that can be used to give a person a virtual sense of being present there.

Other questions stem from a key difference between the solitary seeing we engage in when we observe our home environment with our own natural vision and the enhanced vision we give ourselves when we view the world from a drone camera or by entering an immersive replica of a far-away environment. The latter still often requires an *intermediary* of a kind that is not necessary when we use our unaided vision.

Even when one removes the human go-between that is essential to recording as speech—the person who records an event in the world and then shares that recording with us—there is still *some* actor, who, in some sense, comes between the perceiver and the world they perceive. These include companies that design the technology that a person uses to see with and whose support may be needed to operate it. Even if I use a drone camera to see a certain environment for myself, without having to wait for a recording from anyone else, I will likely have to use a drone and a camera designed by someone else—and receive a video feed or recording over a communication path managed by someone else. The same is true of receiving images I need to immerse myself in a far-away location: I will likely rely on someone else to provide and operate the technology that makes this possible. In most cases that someone else will be a company of some sort. As I have written before, many individuals use Google Streetview or some other "virtual globe" services to reproduce—on their computer screens—the sense of walking down a street in a city thousands of miles away. In a sense, this gives them a power to see such an environment for themselves without having to rely on another individual's (possibly biased and selective) descriptions of it or their (possibly edited) video clip of it. This technology for seeing at a distance gives me a kind of self-reliance that I don't otherwise have. But it isn't a pure form of self-reliance because I am relying on Google or another company to gather and assemble the images—and operate the service—that empowers me to experience such virtual visits.[66]

How then should a framework of rights for extended seeing address the role of such intermediaries? If I have a right to see with technology

[66] Blitz, *Right to Map, supra* note 2.

that extends to use of virtual globes or telepresent experiences, do the companies that design and operate this technology have a derivative right to provide me with that opportunity? If so, does that right nonetheless leave room for government to constrain the way they enable seeing at a distance to ensure that seeing is accurate or, more generally, doesn't undermine the autonomy I am counting on the service to provide? After all, when intermediaries have access to the channels through which we perceive the world, they can conceivably obtain power to shape—and distort—the way we see the world. They might also threaten our privacy and the privacy of those who are in the field of view of a camera or other technology that extends our vision. These kinds of concerns could become even more acute when the intermediary that enables us to alter our perception controls a wearable device that changes how and what we see.

How then should we deal with technologies for extending *solitary* perception and the value we might find in it? As it turns out, there is another part of First Amendment speech protection that might help point a way to move beyond the perception-as-speech-creation paradigm set out in the right-to-record case law. The First Amendment already protects solitary speech—for example, the speech in a diary entry we never intend to share with anyone. As Martin Redish has pointed out, it is hard to justify such protection unless we view the First Amendment as protecting not only communication, but the *private thought* that precedes it and sometimes occurs apart from it. Protection for journal entries, Redish emphasizes, follows from the First Amendment's "commitment to freedom of thought."[67] Kent Greenawalt makes a similar point: The First Amendment, he writes, includes a "principle of freedom of thought" that is "more fundamental even than a principle of speech."[68] It is for that reason that the First Amendment shields not only communication but also "self-communication."[69] It protects our internal reflections. Perhaps then the solitary perception we engage in with technology—the observing that a person can do by themselves, by receiving video or other sensory information, from a machine rather than a speaker—can similarly fall within

---

[67] *See* Redish, *supra* note 14.
[68] R. Kent Greenawalt, *Speech, Crime, and the Use of Language* 46 (1989).
[69] *Id.*

the coverage of freedom of *thought* rather than freedom of expression. It is that possibility that I begin to explore in the next chapter.

REFERENCES

ARTICLES

Marc Jonathan Blitz, *Video Surveillance and the Constitution of Public Space: Fitting the Fourth Amendment to a World that Tracks Image and Identity*, 82 Tex. L. Rev. 1349–1481 (2004).

Marc Jonathan Blitz, *The Right to Map (and Avoid Being Mapped): Reconceiving First Amendment Protection for Information-Gathering in the Age of Google Earth*, 14 Colum. Sci. & Tech. L. Rev. 115 (2013).

Marc Jonathan Blitz and Woodrow Barfield, *Memory Enhancement and Brain-Computer Interface Devices: Technological Possibilities and Constitutional Challenges*, in ed. Veljko Dubljevic and Allen Coin, Policy, Identity, and Neurotechnology (Springer 2023).

Tyler Burge, Perception: First Form of Mind: (Oxford University Press 2012).

Alan K. Chen & Justin Marceau, *Truth and Transparency: Undercover Investigations in the Twenty-First Century* 168 (Cambridge University Press 2023).

Bobby Chesney & Danielle Citron, *Deep Fakes: A Looming Challenge for Privacy, Democracy, and National Security*, 107 Cal. L. Rev. 1753, 1759 (2019).

Andrew Guthrie Ferguson, *Persistent Surveillance*, 74 Ala. L. Rev. 1, 7 (2022).

David Gray & Danielle Citron, *The Right to Quantitative Privacy*, 98 Minn. L. Rev. 62, 65 (2013).

R. Kent Greenawalt, *Speech, Crime, and the Use of Language* 46 (1989).

Sin Duc Hoang, et al., *Harnessing the Power of Virtual Reality: Enhancing Telepresence and Inspiring Sustainable Travel Intentions in the Tourism Industry*, 75 Technology in Society, Nov. 2023 (exploring how "investigates how virtual reality (VR) can enhance telepresence").

Martin H. Redish, *Freedom of Thought as Freedom of Expression: Hate Crime Sentencing Enhancement and First Amendment Theory*, 11 Criminal Justice Ethics (1992).

Christopher Slobogin, *Public Privacy: Camera Surveillance of Public Places and the Right to Anonymity*, 72 Miss. L.J. 213, 236 n.106 (2002).

Christopher Slobogin, *Privacy at Risk: The Government Surveillance and the Fourth Amendment* (2007).

CASES

ACLU of Ill. v. Alvarez, 679 F.3d 583 (7th Cir. 2012).

Anderson v. City of Hermosa Beach, 621 F.3d 1051 (9th Cir.2010).
Animal Legal Defense Fund v. Wasden, 878 F.3d 1184 (9th Cir. 2018).
Askins v. U.S. Dep't of Homeland Sec., 899 F.3d 1035 (9th Cir. 2018).
Berger v. New York, 388 U.S. 41 (1967).
Brief of the Appellant (Louis Zemel), Zemel v. Rusk, 1964 WL 81305 (U.S., 2004).
Buckley v. Valeo, 424 U.S. 1 (1976).
Burstyn v. Wilson, 343 U.S. 495 (1952).
California v. Greenwood, 486 U.S. 35 (1988).
Carpenter v. United States, 585 U.S. 296 (2018).
California v. Ciraolo, 476 U.S. 207 (1986).
Denver Area Educational Telecommunication Consortium Inc. v. FCC, 518 U.S. 727 (1996).
Fields v. City of Philadelphia, 862 F.3d 353, 360 (3d Cir. 2017).
Florida v. Riley, 488 U.S. 445 (1989).
Fordyce v. City of Seattle, 55 F.3d 436 (1997).
Griswold v. Connecticut, 381 U.S. 479 (1965).
Glik v. Cunniffe, 655 F.3d 78 (1st Cir. 2011).
Irizarry v. Yehia, 38 F.4th 1282 (10th Cir. 2022).
Jones v. Opelika, 316 U.S. 584 (1942).
Kaplan v. California, 413 U.S. 115 (1973).
Kleindienst v. Mandel, 408 U.S. 753(1972).
Kyllo v. United States, 533 U.S. 27 (2001).
Lamont v. Postmaster Gen., 381 US 301,308 (1965).
National Press Photographers Ass'n v. McCraw, 90 F.4th 770 (5th Cir. 2024).
Perry Educ. Ass'n v. Perry Local Educators' Ass'n, 460 U.S. 37 (1983).
Planned Parenthood Fed'n of Am., Inc. v. Newman, 51 F.4th 1125 (9th Cir. 2022).
Terry v. Ohio, 392 U.S. 1(1968).
Turner v. Lieutenant Driver, 848 F.3d 678 (5th Cir. 2017).
United States v. White, 401 U.S. 745 (1971).
Va. Pharmacy Bd. v. Va. Consumer Council, 425 U.S. 748 (1976).
Zemel v. Rusk, 381 U.S. 1 (1965).

CHAPTER 3

Freedom of Thought and Revisiting the Right to Receive Information (with Technology)

Abstract If, as the end of Chapter 2 suggested, a right to perceive with technology is often better understood as stemming from our freedom of thought than of freedom of speech—what doctrines can mark the limits of such a right? How can they avoid the problem that led the Court, in *Zemel v. Rusk*, to emphasize that we don't have an unconstrained right to seek knowledge anywhere it is to be found? This chapter explores a number of possible answers to this question—drawing principally from First Amendment doctrines that bar government action when it is (1) motivated by an interest in suppressing speech, (2) attacks speech equivalents, or social practices that have First Amendment value. There can, this chapter argues, be equivalents for these in a doctrine of freedom of thought and perception. But it is also necessary to move beyond these familiar First Amendment models and explore new doctrine for newly defined rights.

Keywords Right to receive information · Right to know · Freedom of thought · Freedom of speech · Constitutional rights · American courts · Social practices · Drone cameras · Coverage

A Right to See Beyond "The Streetlight"

Imagine that the state prohibits us from observing our immediate environment—or from using technology, perhaps with the cooperation of others, to observe remote environments on a video screen. Does any constitutional right prevent it from doing so? If so, what constitutional doctrine in American law raises a barrier against this kind of interference with our perception?

A well-known modern parable can help explain the approach I will take in the remaining chapters of this book to answering such questions. In this parable, a person is engaged in a nighttime search for a set of keys he has lost. He focuses his search within the area underneath a streetlight and soon recruits the help of others to find his keys there. When they think to ask him if this is the area where he lost his keys, he answers "no." The keys were lost in a far-away park but, because he can't see anything in the darkness there, he is instead looking here, where there is light. This tendency is often called "the streetlight effect."[1] We look where it is easiest to see rather in places where observation is more difficult—even if it is the latter hard-to-search area that we need to search in if we are to have any chance of locating what we seek.

Those who seek protection for a right to perception arguably make a similar error when they seek to ground it solely in some connection to speech. The realm of speech has long been shielded by First Amendment doctrine. The Constitution's text expressly protects the "freedom of speech."[2] It says nothing about freedom of perception. Moreover, American courts have built an elaborate jurisprudence to analyze restrictions on speech. There is no equivalent jurisprudence on our right to see or sense. So it may seem to make most sense to try to protect our freedom to observe by locating where judicial light already shines.

---

[1] *See* David H. Freedman, *Why Scientific Studies Are so Often Wrong: The Streetlight Effect*, Discover Magazine, Dec. 9, 2010 [updated May 20, 2025], at https://www.discovermagazine.com/the-sciences/why-scientific-studies-are-so-often-wrong-the-streetlight-effect.

[2] The text of the First Amendment reads: "Congress shall make no law respecting an establishment of religion, or prohibiting the free exercise thereof; or abridging the freedom of speech, or of the press; or the right of the people peaceably to assemble, and to petition the government for a redress of grievances." U.S. Const. amend. I.

In fairness to courts that have decided the right to record cases, the way the law works has given them little alternative: In asking if the Constitution protects recording, they have asked whether it entails creating speech because that is the claim parties have made in the litigation before them. Moreover, the analogy with the "streetlight effect" story is an imperfect one: While looking for keys very far from where they have been lost is pointless, focusing on speech rights in thinking about recording is not. The exchange of photographs and audiovisual recordings one finds on social media certainly *does* extend our perception in a sense: It lets us directly observe events we could otherwise never see or hear. And so protecting speech offers some meaningful protection for our use of technologies that allow us to see and learn from the visual records others communicate to us.

However, this book's mission—to better understand whether there is a constitutional right to see the world with technology, to use bionic eyes, high-tech cameras, video feeds, and extended reality technology—can't make sufficient headway unless we move beyond where the judicial light on this topic is shining most brightly. Rights for speech aren't an adequate foundation for a right to see, and we therefore have to look beyond those rights to find one. That is this book's central argument.

So if we are going to look beyond the First Amendment space already illuminated by case law on the right to record to other parts of constitutional jurisprudence or analysis that might provide a foundation for a right (or perhaps, separate rights) to see with technology, what less fully explored constitutional territory does it make sense to investigate? What other constitutional liberty might our right to perceive the world be a component of or complement to?

In this chapter, and in Chapters 4, and 5, I will explore three separate answers to this question—but all of them, in different ways, make a similar move away from an approach that relies exclusively on free speech protection to shield extended perception. More specifically, all of them move toward considering how speech can be a part of freedom to use our mind (or the biology that underlies it). This chapter begins with the most modest move away from free speech jurisprudence. It will look at arguments that information acquisition or knowledge creation, even outside of any process of communication, is protected by the right to freedom of thought that American courts and legal scholars have said is implicit in the First Amendment. Chapter 4 will then turn from First Amendment law to a different kind of constitutional protection: The protection for personal autonomy in the Fifth and Fourteenth Amendments that protects our right against interference with our bodies and perhaps also our minds. Finally, Chapter 5 will more carefully explore how a *broader* right to mental self-determination or autonomy—independent from the rights to bodily and mental integrity explored in Chapter 4— entail rights to perceive the world with technology.

The Right to Know—And the Right to Observe

The previous chapter, Chapter 2, explained that, although American courts have rightfully described and protected modern image capture as a kind of speech creation, this is at best an incomplete foundation for the variety of ways that we perceive the world, with and without technology. It is hard to categorize as "speech creation" many of the ways we observe the world—with unaided vision, with telescopes, by watching livestream transmissions from a camera mounted on a drone or tower. We're not saying anything when we do nothing but view the world in that way. We are not, simply by observing, creating a digital or documentary record we can share with others. Such observation certainly has value for later communication. It gives us experiences we can then speak about. But this potential for later communication of what we see hardly exhausts the value of seeing or sensing in other ways. There is value for us in learning new information about the world and having the perceptual experience even if we never share it.

What constitutional grounding might there be then for a broader right to see—and to see with technology? I ended the last chapter by suggesting we might find it in freedom of thought rather than freedom of expression. Even when we lack an interlocutor, the First Amendment still protects the way we use language to shape our own minds. It protects our private diary entries, for example, or a painting we never share with the world. It might also, I have argued in the past, protect solitary exploration of virtual reality (VR) environments for the same reason: When we use VR to give vivid form to something in our imagination, we are engaged in something that should be considered an exercise of our freedom of thought. It occurs, I argued, in a "private perceptual realm where, as in a dream or daydream, the government generally lacks any authority to issue commands."[3] Like

---

[3] Marc Jonathan Blitz, *The Freedom of 3D Thought: The First Amendment in Virtual Reality*, 30 Cardozo L. Rev. 1141, 1154 (2008).

other acts of private artistic or literary expression, navigating a VR world we have constructed might give us "greater control over [our] internal mental life" by allowing us to externalize it and better observe it.[4]

A similar logic might extend to private perception. Consider the way that image capture or video recording might be a part of lifelogging. The latter practice, Anita Allen explains, entails creating a "comprehensive archive of an individual's quotidian existence, created with the help of pervasive computing technologies."[5] A person creating such a lifelog might "record and store everyday conversations, actions, and experiences of their users, enabling future replay and aiding remembrance."[6] They might do so, for example, with a tiny camera they wear on a visor, their shirt, or a vehicle they drive—one which captures every action they take and stores in a computer they can access at any point in the future. This kind of pervasive video recording of one's life can certainly provide raw material for communication. One could share clips with friends and relatives. In fact, when Steve Mann pioneered and explained the development of lifelogging with wearable computing, he emphasized the benefits of this kind of "shared visual memory:" If I placed a lifelog on the web, "friends and relatives with wearable Web browsers can see where I have been and catch up with me."[7] However, a lifelog can serve as a "visual memory prosthetic" whether it is shared or not. A key function of this technology then is to support thought and self-understanding. Perhaps certain technologies of extended perception can receive First Amendment protection on that basis.

How might legal scholarship and judicial opinions elaborate such a freedom-thought-based protection for a right to perceive with technology? How can they do so without confronting the same problem that led the Supreme Court to stress, in *Zemel v. Rusk*, that there cannot be an "unrestrained right to gather information"? After all, an unrestrained right to take any action that is an input *to*, or begins *with*, thinking would also be a right that seems all-encompassing: Every intentional act involves some kind of thought (the forming of an intention). Almost every such

---

[4] *Id.* at 1172–1173.

[5] Anita L. Allen, *Dredging Up the Past: Lifelogging, Memory and Surveillance*, 75 U. Chi. L. Rev. 47, 49 (2008).

[6] *Id.*

[7] Steve Mann, *Wearable Computing: A First Step Toward Personal Imaging*, Computer 30(2), Feb. 1997: 25–37, 28.

action adds to our store of memories and thus shapes our mental operations. The same is true even if we focus only on the thought that stems from, draws upon, or shapes perception. If there is a need for some kinds of *outer boundary* on such a right—some kind of *limit* on it—where can it come from?

This chapter will explore two possible bases for limiting a freedom-of-thought-based right to see or sense with technology. Both of these require a digression from the book's focus on seeing and other perception, since they extend constitutional protection to such perception not in its own right but only on the ground that it is a crucial means of acquiring information. One basis for limiting such a right is to restrict only state action that interferes with our information gathering, knowledge creation, or perception more generally when government does so with *motive, purpose, or interest* of preventing us from having certain thoughts or gaining certain knowledge. Government, in other words, may restrict the steps we take to observe the world, free from First Amendment limits, when it does so in the course of taking reasonable measures to protect our health or safety, or our protect the integrity and fairness of financial transactions, or in other regulations that don't involve targeting what we think or the thoughts we express. When its purpose or interest is instead to *target* what we know or think, then—even if it is regulating only information gathering and not communication—it is subject to First Amendment rights-based limits designed to insulate our thought and perception against government control.

A second possible limit lies in defining the scope of the right to see as protecting only *certain* types of perceptions or thinking or certain *social practices* that support perception, memory, and autonomous thought. On this approach, even when government is *not targeting* our perception or our thought—even when the burden or restriction it imposes on them is only incidental to government action with another aim, such as protecting safety—it might *still* cross a constitutional "red line" of sorts, and trigger heightened judicial skepticism, when it enters a "sphere of our lives" that, according to the Supreme Court, is reserved for "an autonomy of self" and must remain strongly insulated against state control. Such "spheres" include not only "freedom" "of expression" but also freedom of "thought" and "belief." It isn't every act we take that involves thought and perception which falls under the shield of such a right. Rather, it is only certain conduct or practices that are central to advancing our knowledge creation or other aspects of mental autonomy.

As explained below, both of these limits have already played a role in the Supreme Court's free speech jurisprudence: That jurisprudence prohibits the government from imposing certain restrictions (1) with the purpose of restricting what ideas people can express or hear and (2) in a way that has the effect of suppressing what counts as First Amendment speech, or undermining a condition of such speech, even when that is not the purpose of the government action. Each of these is true at least where the government action isn't "narrowly tailored" to achieving a "compelling" or "significant" interest—requirements that prevent it from imposing substantially more harm than necessary to our expressive freedom.

It is possible that similar principles can also provide a basis for limiting principles for a right to see with technology, even when such a right is not being exercised as a component of communicating. In previous work, I have explored how each of these principles might play a role in defining the contours of a right to *think or enhance our minds* with technology. Such a right, I wrote, might protect (1) only against government action that has the purpose of preventing us from generating thoughts of a certain kind of shaping our minds in certain ways or (2) more broadly against government action that has the *effect* of disabling us from having certain kinds of mental states or building certain mental capacities or tendencies, at least in cases where the government's measures are not narrowly tailored to achieving significant interests.[8] Perhaps the same principles might provide limits for not only a right to think with technology but also a freedom-of-thought-based right to see with it. In this chapter, however, I will argue that while these limits provide us with a starting point in defining the contours of such a right, we ultimately have to go a further distance from existing free speech doctrine—and the model it provides—to understand how the Constitution applies and should apply to emerging technologies for enhancing perception.

---

[8] *See* Marc Jonathan Blitz, *Freedom of Thought for the Extended Mind: Cognitive Enhancement and the Constitution*, 2010 Wis. L. Rev. 1049, 1084–1090 (2010).

Placing Limits on a Freedom-of-Thought-Based Account of Perception

This chapter turns from speech back to knowledge acquisition and information flow. I say it "turn[s] back" to these ideas because I already discussed, early in the previous chapter, some cases on the "right to receive information" that American courts have found in the speech clause. As noted there, in Zemel v. Rusk, the Supreme Court squarely rejected the idea that there is an "unrestrained right to gather information" entirely disconnected from a link to speech.[9] Receipt of information is protected when we seek to obtain it by listening to a speaker, reading a book, or viewing a diagram. When we obtain it not by serving as an audience for speech in this way but instead by digging for fossils, dissecting an animal, or seeing what we can learn from the experience of cutting timber, then we are still obtaining information—perhaps very valuable information—but are not (except, as noted below, when we are watching a trial or other government proceeding) within the realm of expression shielded by First Amendment law.

In short, the Court generally divides information gathering into two categories: (1) Information received through communication or from expression in some other way and (2) information obtained by individuals engaged in *non-speech* conduct.[10] The first of these is generally protected by free speech law. The second generally isn't. The problem with protecting our silent, non-communicative observing (with or without technology) as a form of information gathering is that it falls most naturally into the second of these two categories, the unprotected one. In the wake of *Zemel*, courts have generally focused on shielding only the first of these two categories. It is also why the information gathering in video or audio recording has been protected not on the ground that the First Amendment allows us to explore and learn about the world, even through non-speech conduct, but rather because such recording is a form of speech creation.

But the freedom-of-thought-based account I am considering here is designed to extend protection to at least *some* non-speech conduct.

---

[9] *Zemel v. Rusk*, 381 U.S. 1, 16–17 (1965).

[10] Barry P. McDonald, *The First Amendment and the Free Flow of Information: Towards a Realistic Right to Gather Information in the Information Age*, 65 Ohio State L.J. 249 (2004).

Imagine, for example, that I fly a drone over my property but aim the cameras it carries at a stretch of public land nearby in order to observe—and perhaps record—wildlife activity in that land. Or perhaps to capture evidence of pollution I believe is occurring in violation of certain laws. Or I might conceivably ask a friend to let me fly such a drone, or fly a drone themselves, over land that *they* own next to such a stretch of public land. It may be true in that situation that I am the *only* observer of the activity on public land.

Observing wildlife on public lands isn't inherently communicative. It may be true, especially if I am making no recording, that the point of my observing is not to communicate what I see to others but rather to learn about it myself. Even so, one might argue, if the First Amendment protects my right to understand and think about the world, then it shields my right to obtain knowledge about it through this kind of observation. Perhaps government officials may still be able to show that there are privacy or security interests at stake that outweigh my First Amendment claim to see and learn about the public land I am observing. But the constitutional value of my perceiving would require that they *demonstrate* to a court that there is a privacy or other interest that overrides those First Amendment interests—instead of simply insisting that they have the leeway to decide, without any judicial oversight, when they can block my observations. Yet if such a right to see—or more broadly, obtain knowledge of the world—isn't limited to observation that is part and parcel of communicative activity, how is it limited? How can an argument for constitutionally protecting it avoid endorsing an "unrestrained right to gather information"?

I noted in Chapter 2 that the Supreme Court in *Zemel v. Rusk* rejected Louis Zemel's argument that he had a right to see Cuba because that argument required that courts recognize an *unlimited* right to obtain information—in any place and through any means a person may use. But visual exploration of one's environment is not the only activity that has confronted the Supreme Court with the challenge of finding limits to the First Amendment's scope. That problem of limits has also arisen in cases on the protection of speech itself. In *United States v. O'Brien*, the Court struggled with the fact that people can seek to convey ideas in innumerable ways. All kinds of wordless conduct could potentially become First Amendment "speech." Someone might express anger at a political decision by throwing a rock or bucket of paint at a government building. If government officials set up a physical barrier intended to block their access to a military base or other sensitive location, a person might signal their opposition to such a limit on their movement by climbing over the barrier or cutting through it. The Court emphasized that non-speech conduct of this sort doesn't become protected speech just because it is intended—by those who take it—to convey an idea. "We cannot accept the view," it said, "that an apparently limitless variety of conduct can be labeled 'speech' whenever the person engaging in the conduct intends thereby to express an idea."[11] So how, we might ask, can the court find a doctrine that says what kind of *limited category of* conduct counts as "speech" and receives the First Amendment protection the Constitution provides for speech? Having asked this question, we can ask if a similar doctrine can provide the basis for a *limited* right to perceive the world—not the "unrestrained right to gather information" the Court rejected in *Zemel*. I will below explore two sets of doctrines from cases on First Amendment free speech law (and a few on access to information) that provide a model or template for a more refined and limited right to receive information. One of these deals with the government's "purposes," "interests," or "motives" underlying the law. A second deals with tests for what is often called "First Amendment coverage" and the question of what kind of conduct that is not intrinsically communicative—or social practice involving such conduct—can count as "speech." Each of these two bases for limiting the scope of First Amendment "speech," as noted above, provides a template of sorts for a right to see with technology rooted in freedom of thought. I will then discuss some of the reasons why it might be good for courts to move a little further from this familiar doctrinal territory, especially as they consider new, less familiar technologies of seeing.

Free Speech, Free Thought, and Government Purposes or Interests

Consider again the reason that the Court rejected Louis Zemel's First Amendment arguments. A core purpose of the First Amendment's free speech protection—perhaps, *the* core purpose of it—is to ensure that individuals can, through democratic discourse, obtain and share the information they need to be informed about the government that answers to

---

[11] United States v. O'Brien, 367 U.S. 376 (1968).

them, and the policy it makes for them, and allow the government to understand and represent their views. However, Louis Zemel stressed in his argument, citizens shouldn't be confined to receiving this information in a second-hand account provided by a reporter or someone else. They should be able to go where the information is to be found and make their *own* observations. To be well equipped to evaluate American policy toward Cuba, for example, it is helpful to learn about that country—and learning about it by traveling there and making direct observations is likely to be far more valuable than simply reading a report issued by the government or a newspaper story.

If the government seems to be restricting information gathering for the *purpose* of restricting the perceiver's ability to learn about the world—rather than to protect others, from say, physical or financial harms that stem from their activity—then that, courts could say, is the kind of government measure that, at the very least, merits judicial skepticism.

The Supreme Court didn't deny that travel to Cuba or other locations had immense educational value for citizens. It simply responded that that point proved too much: Many *non-expressive* actions have the same value so regulating any of them could be described as resulting in "decreased data flow." Driving a car at fast speeds or piloting an airplane can provide insights important to evaluating transportation policy and rules regulating vehicle safety. Experimenting with growing different crops will provide knowledge of farming techniques. That can't mean that all such *non-speech* activities gain a place in the realm of constitutionally protected speech—and a strong shield against government regulation. This First Amendment shield, one might argue, must cover not *all* information-generating activity but only information gathering *through speech*.

But there is an alternative conception of First Amendment protection that broadens it to cover information gathering *without* making it unlimited. On this view, courts should not simply focus on the nature of the activity subject to a government measure and ask if it is "speech" or "non-speech conduct" under the First Amendment. Rather, they should ask *why* the government is restricting it (or *what aspect* of this activity the government is restricting). Such an approach focuses, in other words, on the government's motive or purpose in restricting the information gathering that it is placing limits on. On this view, even if the government is regulating *non-speech* conduct, it may still violate the First Amendment when it is doing so in order to thwart *speech* that this conduct is intended to enable.

As the Tenth Circuit Court of Appeals recently pointed out, that doctrine *could have* provided First Amendment protection to Louis Zemel if he had evidence that the State Department's design in denying his request to visit Cuba was to ensure he couldn't then speak about what he saw there. In that case, the State Department's measure would raise red flags under the First Amendment. This would not be because Zemel's activity is inherently speech-like—and, for that reason, placed by the First Amendment beyond government restriction. Traveling to a foreign country to see it, the Court said, generally is *not* covered by the First Amendment. What links a restriction of traveling to speech is not the nature of traveling but rather the government's censorious purpose in restricting it.

If, in *Zemel*, the State Department had "implement[ed] a law banning travel to Cuba *for the purpose of writing about or filming what they observe*," such a purpose, said the Tenth Circuit, thus would have made *Zemel* a different case—one where First Amendment protection was justified.[12] This might have been clear, for example, if it had previously allowed other Americans to travel there (perhaps on the condition they refrain from writing about it) and said, in internal communications within the Department, that Zemel's request should be rejected so that he cannot, upon returning from Cuba, share his observations with the American public.[13] In this scenario, Zemel wouldn't need to invoke an "unrestrained right to gather information." He would only need to invoke a right against restrictions on information gathering imposed by officials for the *purpose* of preventing the speech it makes possible.

Some well-known and significant First Amendment doctrine is already built around this idea. Consider the doctrinal rule the Court adopted in *United States v. O'Brien* for when the government can restrict or punish expressive conduct—that is a person's expression of a message by doing something to convey that message symbolically rather than by putting in words.[14] In that case, David Paul O'Brien had been arrested for burning

---

[12] W. Watersheds Project v. Michael, 869 F.3d 1189, 1197 (10th Cir. 2017) (emphasis added).

[13] Zemel v. Rusk, 381 U.S. 1, 15 (1965).

[14] United States v. O'Brien, 391 US. 367, 377 (1968).

a draft card to protest the Vietnam War. In doing so, he violated a federal law that forbade destroying draft cards and was prosecuted and convicted for his actions. To determine if it was permissible under the First Amendment for the United States to punish O'Brien for burning his draft card, the Court addressed a number of questions.[15] One of the key questions was about the government interest underlying the law.[16] Was the United States' interest, more specifically, in punishing O'Brien for the anti-war message he had expressed when he burned his draft card? If so, it was engaged in impermissible targeting of ideas. If, on the other hand, the US government was instead simply trying to make sure that draft card records remained available so the military draft could work smoothly, then it was furthering "a governmental interest *unrelated* to the suppression of free expression."[17] Although the Court explained that the First Amendment limits *even the incidental harm* government measures cause to speech in the course of pursuing a *legitimate* government interest, a law faces a much higher First Amendment hurdle—in fact, a nearly insuperable one—when the harm to speech is part of the law's design and not simply an unfortunate side effect. When silencing a particular set of ideas is precisely the law's purpose, it is unconstitutional in all but the rarest circumstances.[18]

The Court's identification of certain interests—in this case, an interest in suppressing expression of certain ideas—as impermissible provides a partial answer to a challenge the Court takes note of elsewhere in *O'Brien*: The challenge of finding principled limits to the First Amendment's staunch protection for speech. As noted earlier, American courts have now recognized that we might speak not merely through verbal conduct in oral assertions or writing. We might also communicate by taking other kinds of physical actions: Protestors might burn effigies, flags, or (as in *O'Brien* itself) draft cards. They might feign death in a "die-in" aimed at protesting a war, or the carcinogenic effects of cigarettes. However, the Supreme Court in *O'Brien* stressed, that does *not* mean that anything a person does to convey a message counts as "speech" shielded by the

---

[15] *Id.*

[16] *Id.*

[17] *Id. (emphasis added).*

[18] *See* United States v. Playboy Entm't Grp., Inc., 529 U.S. 803, 813 (2000) (content-based restrictions on speech are generally subject to strict scrutiny and to survive strict scrutiny, a law must be "narrowly tailored to promote a compelling government interest").

First Amendment. So what does count as First Amendment speech? The Court, a few years later, offered a doctrinal test (the "Spence test") for limiting what kind of conduct intended to convey an idea can count as speech: It must be conduct that the speaker not only takes to convey a "particularized message" but is likely to be understood as such by its audience given the context of the conduct.[19] I will say more about the Spence test later in this chapter.

However, in *O'Brien* itself, its discussion of permissible and impermissible interests provided another such limit: The government only faces something close to an absolute First Amendment barrier in restricting conduct when such a restriction works to advance an interest in the "suppression of free expression." This kind of purpose-based account leaves the government with *some* room to regulate the non-speech conduct that individuals take to convey anger, protest, or other thoughts and feelings. But it simultaneously makes it clear to officials that this leeway is not a blank check that they can use to target ideas they oppose. They could not, for example, constitutionally punish draft card burning *only* when it is done to protest a war. Doing so would clearly indicate the prohibition is meant to punish the protestors' messages and not protect the integrity of the draft card system (which is just as threatened by draft card burning that is done for another, non-ideological reason).

The Supreme Court relied upon a similar principle in a 1992 case, *R.A.V. v. St. Paul*.[20] A city law, it said, could permissibly punish speech that constituted "fighting words"—that is, words that "by their very utterance, inflict injury or tend to incite an immediate breach of the peace"[21]—just as it could punish threats of violence[22] or defamatory words.[23] These kinds of injurious, threatening, or defamatory words involve "speech" in a colloquial sense of that word. But they are not speech protected by the First Amendment. In this sense, said the Court,

---

[19] Spence v. Washington, 418 U.S. 405, 410–411 (1974).

[20] R.A.V. v. City of St. Paul, 505 U.S. 377, 384–90 (1992).

[21] *See* id.; *See also* Chaplinsky v. New Hampshire, 315 U.S. 568, 572 (1942).

[22] *See* Virginia v. Black, 538 U.S. 343, 359 (2003).

[23] *See* United States v. Stevens, 559 U.S. 460, 468 (2010) (noting the First Amendment has historically not protected certain categories of speech content including obscenity, defamation, fraud, incitement, and speech integral to criminal conduct).

fighting words, threats of violence, and defamation are more like non-speech conduct. A set of fighting words is like a punch or other violent action that happens to take verbal form. While threats and defamatory statements communicate, they do so in ways that cause harm—respectively through intimidation or damage to reputation—that perhaps explains why they have traditionally been outside the First Amendment's scope. So a city may restrict them without facing a First Amendment barrier. What a city (or other government) may *not* do, the Court said, is selectively punish *only* those fighting words or other unprotected words that express certain ideas which the city opposes (such as ideas expressing hatred toward certain racial, ethnic, or religious groups, which, in American constitutional law, are protected by the First Amendment). It may not, said the Court, punish fighting words, threats, or defamation only in a way that punishes the government's ideological opponents.

There are a number of accounts we might give of why an otherwise permissible government measure can become constitutionally impermissible when carried out with the wrong purpose—of why, for example, a law that stops travel to a certain location can become impermissible when its purpose is to stop someone from speaking about what they see. One such account emphasizes that just as the morality of an action is often viewed as depending on its motive or purpose—an intentional killing has an entirely different moral status than an accidental one—so government action that is aimed at undermining a crucial interest of ours (such as our interest in speaking freely) should have a different constitutional status than identical government action that has the purpose of safeguarding or advancing our physical welfare. A variant of this might emphasize that, in free, democratic, rights-based societies, the government is expected only to act in ways that respect the rights and interests of the individuals it represents. To act with the purpose of silencing such individuals is to betray this charge, and such a betrayal is constitutionally (as well as morally) problematic even if the government might have acted consistently with its central mission had it done the same thing for a different reason.

One might also justify such constitutional limits on permissible government purposes in terms of the predicted *effects* that flow from purposeful censorship (even when it attacks speech only indirectly by attacking the non-speech conduct that supports or enables it): In short, where government officials restrict what we can observe and what information we can gather *in order to silence subsequent speech about it,* such a systematic attempt at censorship is far more likely to silence us than incidental effects on our speech resulting from other kinds of regulation. A government that bars us from traveling to another country to stop us from communicating about it will likely also take other measures within its power to make such communication impossible. A government that bars exactly the same kind of travel to the same country because it is dangerous will raise no objection if we find safer ways to obtain the same information (for example, by arranging to receive transmissions from cameras located in that country).

However, if the doctrine is sensible, it is also likely to be narrow. It almost certainly does *not* stop the government from barring seeing and any technological enhancement of our perception on the basis of even the slightest risk to privacy. If the government barred creators of mapping programs like Google Streetview from capturing still images of residential streets, for example, it might identify its purpose as protecting whatever risks to privacy arise when Google Streetview does so. Such a privacy-protecting interest is unlikely to be viewed as an interest in what the *O'Brien* court called "the suppression of free expression."[24] The Supreme Court has, in other contexts, said that, when government's interest is in protecting residential privacy that should not be understood as an interest in suppressing speech.[25] A federal appellate court made a similar point in allowing Texas to restrict photography or filming of private property from drone-mounted cameras: "[T]he government," it said, "has a substantial interest in protecting the privacy rights of its citizens" and "drones have singular potential to help individuals invade the privacy rights of others because they are small, silent, and able to capture images from angles and altitudes no ordinary photographer, snoop, or voyeur would be able to reach."[26]

There is, however, a broader version of the government purposes account that might provide more robust protection for perception. This is a freedom-of-thought-based variation of the free speech doctrine that

---

[24] United States v. O'Brien, 391 US. 367, 377 (1968).

[25] It did not view it that way in *Frisby v. Schultz* where it said, in upholding a town ordinance that banned picketing in front of a residential home, that the "State's interest in protecting the well-being, tranquility, and privacy of the home is certainly of the highest order in a free and civilized society." Frisby v. Schultz, 487 U.S. 474, 484 (1988) (internal citation omitted).

[26] National Press Photographers Ass'n v. McCraw, 90 F.4th 770, 794 (5th Cir. 2024).

I have just described: Under the latter doctrine, I have noted, where the government's motive, purpose, or interest in limiting an expressive act is not to counter the physical, economic, or other non-expressive harms that accompany speech, but to crush the speech itself, the First Amendment stands in the way. Some court decisions or scholarly arguments extend this "government purpose" doctrine to a broader set of government purposes: Government acts with impermissible purposes, they argue, not only when its purpose is to stop us from *saying* certain things, but also when its purpose is to stop us from *learning, believing, perceiving, or thinking* about those things.

Under this framework, the Constitution generally allows the government to regulate where we fly a plane or a boat for safety reasons, but *not* for the purpose of keeping us ignorant of what we might see from that vantage point. When the government *aims* to keep us from knowing something we could learn from data or observation, such an aim should trigger First Amendment scrutiny that is absent when it is regulating to address other effects of our physical activity.

Courts have used this framework in the freedom-of-thought jurisprudence. In its most well-known case on freedom of thought, *Stanley v. Georgia*, the Supreme Court seemed to adopt such an approach: Legislation that has the effect of limiting thought is problematic when it is "premis[ed] . . . on the desirability of controlling a person's private thoughts."[27] Applying this precedent, the Seventh Circuit Court of Appeals stressed that since "thought and action are intimately entwined" all laws restricting the latter also necessarily restrict the former—they incidentally place limits on thought. For example, laws that forbid us from trespassing into certain lands prevent us from seeing those lands—and that limits what we can perceive. The First Amendment, it concluded, is therefore violated only by "governmental regulations *aimed* at *mere* thought."[28]

The Seventh Circuit's analysis focused on a claim by a person previously convicted of sexual offenses against children that, when a city banned him from entering public parks where children were often present, they were "punishing him for his private thoughts"—namely, the sexual

---

[27] Stanley v. Georgia, 394 U.S. 557, 566 (1969).

[28] Doe v. City of Lafayette, Ind., 377 F.3d 757, 765 (7th Cir. 2004)(emphasis added to "aimed").

thoughts he had regarding children while in the park. The Seventh Circuit emphasized two points in rejecting his claim. First, it noted, the city's prohibition wasn't restricting "mere thought, unaccompanied by conduct." It was rather prohibiting him from *physically locating* himself in a particular place and watching children present there. That, his action involved "thought *plus conduct*."[29] Second, the city's bar on his entering a park was restricting the *conduct* component of that combination: The city, said the judges, "ha[d] not banned him from having sexual fantasies about children." Rather it barred him from taking steps to locate himself near children, where it was far more likely he would act on those thoughts.[30]

This government purposes framework might, as I have argued in earlier work, also be applied to how the government regulates our use of technology to shape our thoughts. Imagine, for example, that a person wishes to reshape the way they think with a brain-computer interface device—such as a brain-controlled video game—that reduces the stress they feel in the face of certain challenges. Or that they wish to increase their ability to focus or remember details by using "cognitive enhancement" technology such as certain drugs or other technologies that might enhance their mental capacities. Or wish to undergo a certain kind of psychotherapy to become more independent—and resistant to pressures to conform to the "groupthink" they believe exist in the communities they are a part of.[31] The government, of course, regulates such drugs, devices, and psychological counseling methods to ensure they are safe for clients—and not marketed to consumers in ways that are misleading. But that doesn't mean that the government may regulate the same tools of shaping our thought in order to prevent us from thinking in certain ways it doesn't want us to think. When the government's *purpose* in regulating our use of neurotechnology or our use of a certain therapy technique is to ensure we continue to think in ways it wants us to think, it is doing something that is constitutionally suspect—it is potentially violating our freedom of thought.[32] That doesn't mean that the government can never premise its action

---

[29] *Id.*

[30] *Id.*

[31] *See* Marc Jonathan Blitz, *Free Speech, Occupational Speech, and Psychotherapy*, 44 Hofstra L. Rev. 681 (2016).

[32] *See* Marc Jonathan Blitz, *Freedom of Thought for the Extended Mind: Cognitive Enhancement and the Constitution*, 2010 Wis. L. Rev. 1049 (2010).

on preventing us from generating certain mental states or tendencies. It might, for example, prohibit the manufacture, sale, or use of a drug that induces severe and lasting depression or psychotic episodes even if certain people wish to take such drugs.[33] But as a general matter, as the Supreme Court emphasized in *Stanley v. Georgia*, the government cannot be in the business of wresting, from individuals, choices over the content and operation of their own minds.

This extension of a government purposes account to the way we think with technology might also cover the way we see with technology. If the government bars driving to a certain location because it doesn't want us to see and learn about illegal pollution or its failure to maintain that land, this may trigger First Amendment scrutiny of a kind that would not apply if it is clearly imposing the prohibition to protect our safety or that of other people. Thus, in earlier work on government restriction of image capture and virtual exploration, I argued that courts should be deeply skeptical of government restrictions on information gathering where there is "reason to suspect that" government is imposing such restrictions "not to block individuals from causing harm as they seek knowledge . . . but rather from the knowledge itself."[34] A restriction on information gathering, in other words, would be entirely permissible under the First Amendment when it is imposed to further "legitimate concerns about how the activity involved in that information gathering might affect the public" but not when it is designed to give force to "a desire on the part of the government to keep people in ignorance."[35] Like a government purpose of "suppressing expression," a government purpose of suppressing knowledge of one's environment should be deeply suspect. In this respect, courts would bring to information gathering an approach similar to the one the Supreme Court brought to symbolic conduct (like burning a draft card to protest a war) in *United States v. O'Brien*. In fact, one of the dissenting opinions in *Zemel v. Rusk*—written by Justice Douglas—suggested that precisely such an *O'Brien*-type

---

[33] *See* Marc Jonathan Blitz, *A Constitutional Right to Thought Enhancing Technology*, in ed. Veljko Dubljevic and Fabric Jotterand, *Cognitive Enhancement: Ethical and Policy Perspectives in International Perspective* (Oxford University Press 2016).

[34] Marc Jonathan Blitz, *The Right to Map (and Avoid Being Mapped): Reconceiving First Amendment Protection for Information-Gathering in the Age of Google Earth*, 14 Colum Sci. & Tech. L. Rev. 115, 188 (2013).

[35] *Id.*

analysis could be applied to Louis Zemel's plan to observe and gather information from Cuba: Just as courts can apply the First Amendment to "speech brigaded with action" to ensure the government is targeting the action and not the speech embodied in it, so it can apply a similar analysis when the information gathering that precedes and enables speech is itself brigaded with action.[36]

Under this approach, the government could not always immediately deny First Amendment status to an individual's interest in observing their environment simply by plausibly invoking others' right to privacy. Protecting such a right to privacy may often not be an interest in "suppressing expression." Preventing someone from observing something that is arguably "private" isn't punishing them for an idea they have communicated. It *is*, however, stopping them from learning and thus knowing about it. Indeed, a person's informational privacy *depends on* preventing others from observing or learning certain things about that person. Privacy protections, I argued in earlier writing, "generally are designed precisely to keep someone from learning something about the world." Some of these privacy laws do so by limiting speech: They prevent individuals from hearing facts conveyed by others. Other privacy laws prevent "uncovering and observing" facts. Either way, I argued, laws should at least face judicial skepticism when "they are designed to stop information from reaching people (and do not simply disrupt such data flow while attacking another harm)."[37]

Jane Bambauer has proposed a similar but broader First Amendment framework. She does not limit such a framework to the perception of, or information gathering in, physical or virtual environments. Rather, she argues that the protection the First Amendment provides for thought and especially to knowledge creation should extend more generally to generating, accessing, and disseminating data.

The First Amendment, she argues, entails a "right to create knowledge" which "ensures that the state will not interfere unduly with its constituents' learning."[38] This right, says Bambauer, supports extending First Amendment protection to disseminating data—and also a First Amendment right to obtain or generate such data. In response to the

---

[36] Zemel v. Rusk, 381 U.S. 1, 15 (1965).

[37] Blitz, *The Right to Map (and Avoid Being Mapped)*, *supra* note 34, at 192.

[38] Jane Bambauer, *Is Data Speech?*, 66 Stan. L. Rev. 57, 87 (2014).

concern that it isn't tenable to think of the First Amendment as protecting *every* action that contributes to knowledge or information gathering, Bambauer's answer is to stress the "manageable limiting principles to the First Amendment's protection of knowledge" that one finds in "purposive" or "motive" analysis of the government's restriction. Many laws that limit knowledge, she points out, do not "hav[e] a direct purpose to obstruct knowledge."[39]

By contrast, she says, First Amendment limits should apply to laws that "ha[ve] the very purpose of limiting knowledge."[40] Under this principle, laws that restrict observation in order to protect privacy should not easily escape judicial scrutiny. Although protecting privacy is unlikely to be understood by courts as suppression of speech, it almost inevitably involves suppressing knowledge. "Data privacy laws," as Bambauer notes, "have the unabashed goal of limiting, and shaping, what the government's constituents can know."[41] This, she says, should subject them to some skepticism from courts, because by limiting knowledge generation, such laws can threaten a key purpose of the First Amendment—namely, to create and generate knowledge.[42]

This does not mean that privacy laws are generally unconstitutional. Government might sometimes have a powerful interest in limiting knowledge to protect our individual privacy. But it can't simply insist that courts trust and defer to its judgment in that case. Given the potential damage it does to knowledge creation any time it restricts our access to data to keep us in ignorance of what the data reveals, courts—on Bambauer's view should instead insist that the government demonstrate the need for such restriction, and show it is not restricting our data access or use more than they need to. Courts should, in other words, apply some form of heightened scrutiny, rooted in First Amendment doctrine, to ensure government's interference with the way we learn about the world is justified by powerful interests (in privacy or some other legitimate reason for restricting information flow).[43]

---

[39] *Id.* at 90.
[40] *Id.* at 87.
[41] *Id.*
[42] *Id.*
[43] *Id.* at 91.

Moreover, Bambauer argues, it is essential to demand such a showing even when privacy is at stake because "data" is often "essential to our thought processes." So even though lawmakers might justifiably protect the "liberty-preserving" functions of privacy—the way it enables us "to act authentically" in ways that would be impossible if we were subject to constant observation—lawmakers should not, in doing so, have free rein to sacrifice the equally important "liberty-preserving" function of "knowledge gathering."[44] She also emphasizes that while some privacy interests, such as the right to seclusion in our homes and private conversations, and the right to privacy in certain relationships (such as those between doctors and patients) clearly have importance,[45] and should likely survive First Amendment challenges, there are also "[s]ome existing privacy laws should not be able to withstand constitutional scrutiny."[46]

This kind of government purpose analysis identifies one important component that should likely be a part of constitutional protection provided by a right to see with technology. Where the government is limiting our ability to see the world *in order to* prevent us from learning something about it, it—at the very least—must justify its restriction under the First Amendment by explaining why its restrictions are necessary to protect privacy interests or some other interests of significant weight.

This purpose-based analysis presents an answer to some of the questions I raised in the last chapter and at the beginning of this one. But such an answer is at best a partial one. In the first place, if our ability to exercise a right to see must generally be balanced against other interests that the public has against our capacity to see certain things without restriction, it is unclear—without contextual inquiry into how such a balance should be struck—what kinds of interactions such a right will and will not protect. It thus lacks the clarity provided—in First Amendment cases on speech—of the high bar that courts raise against intentional measures the government takes to suppress our *expression* of certain ideas. Preventing us

---

[44] Jane Bambauer, *Is Data Speech?* at 105.

[45] *Id*. at 111–112.

[46] In contrast to my account, which suggests applying the O'Brien test to laws designed to limit observing or information-gathering, Bambauer argues that the level of scrutiny should "depend on context," *Id*. at 105, and stresses that "privacy problems can be addressed through laws that are narrowly tailored to compelling interests in seclusion and confidentiality and thus survive scrutiny at any level." *Id*. at 110.

from *learning* about certain things won't face the same nearly insuperable constitutional hurdle.

Consider again my earlier argument that government interference in our observation of our environment might be subjected to a variation of framework from *United States v. O'Brien*—wherein it faces greater judicial scrutiny not only (1) when the motive that drives its action, or the interest that underlies it, is the "suppression" of "expression" but *also* (2) when it is "the suppression" of *knowledge*—"a desire on the part of the government to keep people in ignorance" or, more generally, "to stop information" from reaching us. The analogy to *O'Brien* I drew there is only a partial one. I suggested a departure of sorts from *O'Brien*— a disanalogy between speech restrictions and restrictions on information gathering. And that departure is a significant one. When legislators enact laws *designed to suppress the expression of certain ideas*, the Court adopts a very strong presumption that this purposeful censorship of expression is unconstitutional. No matter how strongly legislators might insist that the ideas they are targeting are noxious or worthy of condemnation, courts almost always find that such purposeful censorship is at odds with the First Amendment. After all, it is not lawmakers' job to judge for us what ideas are unworthy of being expressed and heard. The First Amendment lets us make that judgment for ourselves. The government's interest in suppressing ideas, in preventing them from being expressed, is generally entitled to no weight—and thus cannot outweigh the freedom we have to express ideas and autonomously evaluate them.[47]

By contrast, on the account I've defended before and am presenting again here, laws designed to prevent us from *observing and recording* private aspects of others' lives should *not* generate the same level of skepticism, even though they are, in doing so, designed to keep us in ignorance of certain (private) facts.

Such privacy-protecting laws are not nearly as likely to be a betrayal of the government's role in a free society as a law that punishes or filters out particular messages. On the contrary, when it comes to privacy law,

---

[47] As explained below, the government can still—in very rare cases—gain court approval to suppress speech on the ground that the dissemination of the information in it is dangerous. But it has to satisfy an extraordinarily challenge test—called "strict scrutiny"— which courts rarely find the government can meet. Moreover, even when strict scrutiny applies, the government cannot satisfy simply by arguing that certain ideas are unworthy of ever being entertained. It must rather point to other some other "compelling interest" related to the harms that certain speech content would very likely cause.

there often is a balance to be struck between two legitimate interests: (1) Our interest in freedom to obtain knowledge about the world and (2) the government's (and the public's) interest in assuring that individuals can retain a sphere of privacy that is essential their freedom. Indeed, privacy is sometimes just as essential a condition for intellectual exploration as an opportunity to observe the world. Consequently, a government interest in preventing an observer from learning about a private matter will not disqualify the law it justifies from being constitutionally permissible in the same way as a government interest in barring expression of particular ideas. That should also be true where a person's accessing of information would threaten personal security or safety in other ways. It is only, in certain circumstances, where privacy or security interests are absent or very weak that a government purpose of suppressing information gathering would merit as much skepticism from courts as a government purpose of suppressing expression of particular ideas.

In short then whereas a law with a motive of suppressing speech will almost automatically be unconstitutional, a law with a motive of preventing perception or acquisition of knowledge will *not* be—if the government is standing in the way of knowledge to shield personal privacy or security. Government officials defending the law will still have to explain why their privacy-protection interests trump our interest in observing and learning about the world. But they will not, in doing so, face the same almost insuperable odds they face when their purpose is crushing expression of certain ideas.

But if that is true, then it is unclear whether a freedom-of-thought-based right to receive information is more extensive or robust than the freedom-of-speech-based right to receive information discussed in Chapter 2. As I noted there, the right to record provides an answer of sorts to how we can reconcile our own interest in augmenting our perception—so that we can directly observe remote events—with others' property and private rights. "Where we reach a limit of our own autonomy," I noted, "we can still share in – through communication of images and videos – the perceptions available in someone else's realm of autonomy." If our own right to explore and learn about the world is necessarily limited by others' privacy rights, how much broader is it than our right to view what others make available to us in the form of video or audio recordings? Still, one might argue, such a right would at least require government to offer a justification—and a justification of a certain strength—to override our perception of an environment, even when that perception is not part of any process in which we receive communications from others.

There is, however, another problem with making the right to see depend solely on government purposes-based account. It is a problem similar to that which plagues using a pure purposes-based account to protect a right to speak or a right to think. The problem is that government officials can do substantial damage to speech, thought, or perception even when they are targeting something else—and that damage is sometimes unnecessary to achieve such a goal even when it is legitimate. Thus, the Supreme Court in *United States v. O'Brien* did not find that government can restrict speech as heavily as it likes so long as its goal is something other than censoring ideas. Even when such restriction is an incidental effect of the government's regulation rather than its aim, the damage it does to expression is still harmful to what the First Amendment is meant to protect—and therefore, the Court has said, is only permissible when the government interest is "substantial" and speech restriction is not "substantially broader" than it needs to be. The same should be true, I have argued, of government regulations of the technologies we use to think with.[48] It should also be true, I argue here, of our right to *see and sense* with technology. But that raises another question: If government can cross a constitutional line by interfering with our perception of the world—with or without technology—and if it can do so even when keeping us in ignorance is *not* its goal, then it is necessary to understand more about what this places off-limits: If government needs a justification to interfere in our seeing or sensing, regardless of the purpose government is acting with, then what realm does this presumptively place off-limits for officials?

SOCIAL PRACTICES AND FIRST AMENDMENT COVERAGE (FOR SPEECH, THOUGHT, AND PERCEPTION)

It is helpful to begin to address this question by looking more closely at speech protection. The First Amendment's protections for "speech" don't only protect our speaking and writing. They also protect the way we express ourselves, as *O'Brien* notes, in symbolic conduct. They also protect the multitude of things we do to generate and disseminate speech:

---

[48] Blitz, *Freedom of Thought for the Extended Mind, supra* note 8, 1089–1090.

The thinking and learning that gives us something to say, the distribution that delivers a completed book or article to its audiences, and the listening and reading that ensures that a speaker can communicate their ideas and is not confined to speaking without any audience. But as broad as First Amendment speech protection is, the Court also stressed in *O'Brien* that it cannot be "limitless."[49] It cannot constrain the government every time it regulates other conduct that could conceivably be used to convey ideas or play a role in creating speech. How then do courts mark the limit of free speech protection? Can the boundaries they draw for that also point the way to drawing boundaries for First Amendment protection for thought and observation?

The question I asked in the previous section of this chapter may also require addressing the question I am raising here: In that section, I considered the possibility that government violates the First Amendment when it acts in order to target speech or—in a variant of that approach—in order to target knowledge creation or other thought. But in order to know when the government is targeting First Amendment-protected speech or thought, we may sometimes have to know what those categories refer to. Imagine, for example, that government bars a company from recording thousands of hours of footage from a fleet of camera-laden drones that automatically record whatever occurs below them in an urban landscape over a period of days. If the government does so because it believes it is wrong for companies to capture and preserve that much information about human activities, is its restriction aimed at thwarting First Amendment "speech"? Is such a fleet of drones engaged in creation of First Amendment speech? Does it matter if there is any plan to share the footage with viewers in the general public? Or imagine that a law bars us from altering certain footage in a lifelog to try to deceive *ourselves* about our *own* past because government believes that doing so distorts our self-understanding and causes psychological harm. Would such a law barring falsification of records of our past (even if we desire it) be targeting the "thought" protected by our freedom of thought?

As it turns out, these are difficult questions. And while difficult, they are probably unavoidable for First Amendment law—not only because they are sometimes necessary to assess whether the government is acting with an impermissible purpose but also because sometimes, those

---

[49] United States v. O'Brien, 391 US at 376.

purposes aren't decisive. First, even when the burdens that the government imposes on speech are only side effects of laws rather than their aim, they still face First Amendment hurdles. It is true that First Amendment law raises a particularly strong barrier against laws the government takes *in order to* suppress the expression of ideas. In the parlance of modern First Amendment jurisprudence, such intentional censorship is unconstitutional except in the very rare cases that the government can overcome "strict scrutiny"—that is, the government must show that its measure is necessary to achieve a "compelling government" interest that cannot be achieved with any less speech-restrictive measure.[50] The First Amendment's protection is generally weaker against measures that *incidentally* limit speech in the course of giving force to a different government purpose—such as a law that *aims* to ensure the physical integrity of draft cards but has the *side effect* of preventing anti-war messages that involve burning them. But while weaker, such protection is not meaningless. Such incidental restriction of speech faces "intermediate scrutiny." The government's legitimate interests need not be the kind of extraordinary "compelling government interest" of the highest order necessary to justify intentional censorship. Yet they must still be weighty enough to count as "substantial" or "significant."[51] Moreover, even incidental restrictions on speech must be "narrowly tailored" to achieving those interests. They need not be the least speech-restrictive means of achieving the government's aims. Yet they must be not "substantially broader" than they need to be. In short, whenever the government restricts or burdens First Amendment speech—whether it does so intentionally or incidentally—it must meet a kind of heightened scrutiny. So it is often important to know when the conduct is restricting or burdening counts as "speech."[52]

---

[50] *See Playboy*, 529 U.S. at 813 (content-based restrictions on speech are generally subject to strict scrutiny and to survive strict scrutiny, a law must be "narrowly tailored to promote a compelling government interest"); The "Government may . . . regulate the content of constitutionally protected speech in order to promote a compelling interest if it chooses the least restrictive means to further the articulated interest." *Sable Commc'ns of California, Inc. v. F.C.C.*, 492 U.S. 115, 126, (1989).

[51] *See* Turner Broad. Sys., Inc. v. F.C.C., 520 U.S. 180, 189 (1997).

[52] *Id.* at 189 ("A content-neutral regulation will be sustained under the First Amendment if it advances important governmental interests unrelated to the suppression of free speech and does not burden substantially more speech than necessary to further those interests").

As noted earlier, the Supreme Court has provided one test for answering this question—the "Spence test"—in the 1974 case of Spence v. Washington. A student in that case was arrested under a Washington State law for hanging an American flag upside down with a peace symbol affixed to it to express distress at the killing occurring during the Vietnam War. The display of the flag, said the Court, counted as First Amendment speech because the student engaged in it to convey a "particularized message" and "in the surrounding circumstances the likelihood was great that the message would be understood by those who viewed it." This then is one doctrinal test that one might use to determine if regulated conduct counts as First Amendment "speech."[53]

Legal scholarship has been quite critical of the *Spence* test. In an influential essay, Robert Post described it as "transparently and manifestly false."[54] This is in part because it excludes from the realm of First Amendment "speech" some conduct that clearly is, such as an avant garde movie with no discernible message.[55] Abstract art of the kind painted by Jackson Pollock lacks any "particularized message." So does the instrumental music composed by Arnold Schoenberg. Yet these works of art are, as the Supreme Court pointed out, "unquestionably shielded by the First Amendment."[56]

The Spence test also makes the opposite error: It not only excludes from the category of First Amendment "speech" some activities, such as abstract art or music, that clearly fall within it. It also includes some activities that courts invariably place outside of it. Throwing a rock through someone's window after they criticize a particular local political figure may clearly convey anger and opposition to their criticism. But that doesn't make it expression protected by the First Amendment.

In fact, many statements that convey a particularized message and that we call "speech" in ordinary parlance are not speech for First Amendment purposes. As Frederick Schauer has pointed out, "'Speech' is what we use to enter into contracts, make wills, sell securities, warrant the quality of

---

[53] Spence v. Washington, 418 U.S. 405, 410–411 (1974).

[54] Robert Post, *Recuperating First Amendment Doctrine*, 47 Stan. L. Rev. 1249, 1252 (1995).

[55] *Id. at 1253*.

[56] Hurley v. Irish-Am. Gay, Lesbian, Bisexual Group of Boston, 515 U.S. 557, 568 (1995).

the goods we sell, fix prices, place bets, bid at auctions, enter into conspiracies, commit blackmail, threaten, give evidence at trials, and do most of the other things that occupy our days and occupy the courts."[57] But regulation of many of those examples of speech is not considered to raise First Amendment questions at all. That some of these forms of speech sometimes meet the Spence test doesn't change matters: A statement in a set of instructions a company provides for how to use a product that it manufactures and sells may well convey to a "particularized message" to purchasers of that product about how to use it safely. That doesn't automatically make it First Amendment speech.

How then can we know which actions we take to communicate or express ourselves count as "speech"—not only under the colloquial definition of "speech" but for First Amendment purposes? Scholars like Post, Schauer, and others argue that this is, at least in large part, a matter of social convention. According to Schauer, the boundaries that define speech may derive less from any "underlying theory of the First Amendment and more from the political, sociological, cultural, historical, psychological, and economic milieu in which the First Amendment exists and out of which it has developed." Post does not leave theory so fully behind. According to him, what the First Amendment covers is not speech per se but rather "discrete forms of social practice that speech makes possible."[58] The reason that courts should focus on such "social practices," Post argues, is that it is these practices—*not* any acts of speaking viewed separately from them—that serve the values the First Amendment is meant to protect. For example, one reason that the Constitution protects speech is to provide a crucial foundation for a well-functioning democracy: Voters cannot choose elected officials and elected officials cannot then represent their constituents unless individuals can have robust debates about policy in which they are free to question, defend, or criticize existing policies and or proposed alternatives. Another core purpose of free speech is to prevent censors from blinding us to the historical or scientific truth by thwarting intellectual exploration and the sharing of information. But, as Post stresses, although "[s]peech is of course prerequisite for both democracy and truth-seeking [] speech

---

[57] Frederick Schauer, *The Boundaries of the First Amendment: A Preliminary Exploration of Constitutional Salience*, 11 Harv. L. Rev. 1765, 1773 (2004).

[58] *Id.* at 1786. Post, *supra* note 54, at 1274.

alone, in the absence of other necessary social practices, will not yield the values we seek in either democracy or truth-seeking." More generally, he writes, "[s]peech does not itself have a general constitutional value, but rather we attribute to speech the constitutional values allocated to the discrete forms of social practice that speech makes possible. The unit of First Amendment analysis, in other words, ought not to be speech, but rather particular forms of social structure."[59] This helps explain why the First Amendment does not shield *every* act in which we use words—or nonverbal conduct—to communicate. Some of those means of expressing ourselves (throwing a rock through a window to convey anger, or bidding at an auction, or putting words in a contract to enter a commitment) don't advance the values the First Amendment exists to advance. Others, such as engaging in debate with fellow citizens about political issues, do.

Might one bring the same type of focus on social practices to a jurisprudence of freedom of thought and perception? There is a sense in which First Amendment law already has. It has protected a certain type of information gathering not solely because, as with the video recording discussed in Chapter 2, it classifies the information gathering as speech creation. Rather, it protects observation of a kind that is connected to the amendment's underlying *purposes* rather than its textual and traditional focus on "speech."

The Supreme Court itself has already made a move of this sort: It has carved out a significant exception to the rule that the First Amendment only protects a right to receive information only from a "willing speaker."[60] In *Richmond Newspapers v. Virginia*, it held that members of the public have a right to observe a criminal trial—not just the verbal statements made in it, but other aspects of it—and that they have a right to do so even when the prosecution, defense, and judge prefer to prevent such observation.[61] The basis for finding that the public had a right to observe this kind of judicial process wasn't then that the public had a right to be an audience for a speaker who wished to deliver information to it. The right to observe a trial is instead rooted in one of the fundamental purposes of the First Amendment—to ensure that citizens can learn about

---

[59] See Post, *supra* note 54, at 1272–1273.

[60] Va. State Bd. of Pharmacy v. Va. Citizens Consumer Counsel, 425 U.S. 748, 756 (1976).

[61] Richmond Newspapers v. Virginia, 448 U.S. 555, 576 (1980).

the way their government functions and safeguards the rule of law—and can keep it accountable.[62] Public access to, and observation of, criminal trials is, in other words, a long-standing social practice that supports critical First Amendment values, even when such access or observation does not involve speaking.

Of course, it is not simply observation of criminal trials that is important for keeping citizens informed about their government's operation. Citizens can learn crucial information about their government's essential operations by observing—and at times, sharing recordings of—the way police officers patrol the streets and interact with individuals there and of whether (and how) various government agencies protect the environment or improve road safety. Not surprisingly, the role that audio and video recordings play in informing citizens about matters of public concern has provided an alternative rationale for giving them First Amendment protection. When the Seventh Circuit Court of Appeals, in *American Civil Liberties Union v. Alvarez*, found that the sound recording component of ACLU's planned video recordings of police merited First Amendment protection, it did so not only because such recordings were a kind of "speech creation" but also because they furthered the same crucial First Amendment purpose that the led the Supreme Court, in *Richmond Newspapers*, to find the public had a right to observe criminal trials. For "the founding generation," it said, "the liberties of speech and press were intimately connected with popular sovereignty and the right of the people to see, examine, and be informed of their government."[63] The recordings the ACLU wished to make of "government officials performing their duties in public" clearly furthered this First Amendment purpose. Note that while this case was specifically about the making of recordings, the above language suggests that the First Amendment's shield can extend more generally to acts taken by Americans to "see, examine, and be informed of their government" in other ways—perhaps even without creating a recording.

We might similarly ask if the First Amendment's protection of thought, and specifically perception, could work in the same way. Is it possible that there are certain social practices that are central to the way we think or the way we observe our environment that merit protection under a

---

[62] *Id.*

[63] *ACLU of Ill. v. Alvarez*, 679 F.3d 583, 599 (7th Cir. 2012).

right to think and perceive with technology, even if there can be no general "unrestrained right to gather information"? I have argued in the past that, at least when it comes to a right to freedom *of thought*, we can identify such social practices. In marking the coverage of freedom of thought, I argued, we might adapt Post's argument against viewing the First Amendment's speech clause as applying to every act of speech. Similarly, we might refrain from assuming that freedom of thought can "cover every external act that enables a subsequent act of perception, imagination, or internal deliberation."[64] For much the same reason that the right to receive information cannot cover every action we take that gives us new experiences and new information, we cannot plausibly understand freedom of thought to "empower us to generate any mental state we wish through any means we choose." Rather, what this right covers are those social practices "that help mark out, and allow us to take advantage of, such a sphere of self-sovereignty" in our intellectual lives. Among such practices are those which Neil Richards has defended under the heading of "intellectual privacy": Reading books, writing in diaries, viewing films, or conducting Internet searches.[65] I have defended First Amendment protections for using libraries on similar grounds.[66] And the staunch protection the Court provided in *Stanley v. Georgia* for our right to shape the cultural life of our home—the personal library of books and films—likewise protected a social practice that is essential to how modern courts understand the right to freedom of thought. All of these social practices, I have argued, are important for our exercise of freedom of thought and not simply because they generate mental states—all action we take does that—but rather because they help maintain "a wall of sorts between our internal life and the external, socially-shared world" and they give us sovereignty over that internal life.[67]

To be sure, there may also be social practices in which we work *with others* to shape our thoughts. Thus, some scholars have defended a right to engage in scientific inquiry and experimentation as an exercise of

---

[64] Blitz, *Freedom of 3D Thought*, *supra*, note 3, at 1198.

[65] *Id.* at 1199. *See* Neil Richards, *Intellectual Privacy: Rethinking Civil Liberties in the Digital Age* 4 (Oxford University Press 2015).

[66] *See generally* Marc Jonathan Blitz, *Constitutional Safeguards for Silent Experiments in Living: Libraries, the Right to Read, and a First Amendment Theory for an Unaccompanied Right to Receive Information*, 74 UMKC L. Rev. 799 (2006).

[67] Blitz, *supra* note 3, *at* 1200.

freedom of thought. Dana Remus, for example, has argued that "any regulation on scientific experimentation will impact scientific thought."[68] Other scholars, such as Natalie Ram and Jane Bambauer, have likewise argued that restrictions on scientific experimentation will interfere with the generation of knowledge.[69] More generally, Simon McCarthy-Jones has argued that while individuals should certainly be able to invoke freedom of thought protections against interference in their solitary thinking and use of thinking tools, because "we reason better together than alone, then maximal freedom of thought is to be found in groups, not individuals."[70] So freedom-of-thought protection must extend to some of the thinking we do collectively—certainly to the indispensable role that our collective discourse plays in shaping our thought (something which, in the United States, is in any case already shielded by First Amendment free speech protections) but also, perhaps, to other collective endeavors, or components of these endeavors, that promote our thinking.

There are, however, at least two problems with understanding freedom-of-thought protection solely, or perhaps even primarily, in terms of protecting certain social practices that support First Amendment values. First, many hypotheticals that are often viewed as paradigmatic interferences into a person's freedom of thought are not most naturally described as interference in any social practice. Imagine, for example, that state officials find a way to generate certain mental dispositions (such as docility and acceptance of government actors' decisions) by inserting chemicals

---

[68] *See* Dana Remus Irwin, *Freedom of Thought: The First Amendment and the Scientific Method*, 2005 Wis. L. Rev. 1479, 1504–1505.

[69] Natalie Ram, *Science as Speech*, 102 Iowa L. Rev. 1187, 1195, 1198 (2017) (noting that "concern for knowledge production is [] at the core of" multiple First Amendment theories and arguing for First Amendment protection for "scientific experimentation" because it "is one of the primary means by which people develop new knowledge"); Jane R. Bambauer, *The Empirical First Amendment*, 78 Ohio St. L.J. 947, 947 (2017) (presenting an argument that "[t]he First Amendment should protect not only the right to share ideas and factual claims, but also a (limited) right to test them" scientifically).

[70] Simon McCarthy-Jones, *Freethinking: Protecting the Freedom of Thought Amidst the New Battle for the Mind* 56 (2023).

into the food ingested by a particular activist.[71] Apart from being a violation of that person's bodily autonomy, such a measure is also intuitively an impermissible type of thought control. Unlike a verbal argument meant to persuade the activist to accept government policies, it is a kind of influence over the person that would likely count—in the words of *Stanley v. Georgia*—as a flatly unconstitutional attempt to control men's minds. However, that is not primarily because such mind control is interfering in any social practice regarding the ingestion of food. It is rather because the state is forcibly and perhaps surreptitiously, entering and exercising control in a realm of individual autonomy where it has no place. The same intuitively seems true if the state is able to exert control over a person's thinking in other ways. Imagine, for example, that the state finds a way to reshape certain unconscious tendencies or preferences of mine. If, for example, the state were able to use "subliminal stimuli" to shape my mental tendencies outside of my awareness, this would intuitively be a violation of my right to mental freedom. But that is not because the state, in doing so, is interfering with anything that can be easily described as a "social practice." The preferences I have unconsciously formed, and may not even know yet that I have, aren't social practices, and the control that I or others exercise over them is not necessarily part of any such practice either. Some paradigmatic violations of mental freedom thus interfere with our thinking—and do so in a way that would be unconstitutional under the First Amendment—but not by interfering with social practices that enable or shape that thinking.

When it comes to a right to freedom of thought, mental determination, or self-determination, in any case, we can intuitively identify violations of those rights in government interferences with our person and private mental activity, even when those interferences can't easily be described as violating a social practice. Of course, the principle of non-interference that a law gives force to is itself largely a product of history and social processes (that create ethical standards). Yet the activity shielded by freedom of thought in these cases is not a discrete social practice. This is perhaps most clearly the case when the right being violated is the kind we might

---

[71] *See* Jan Christoph Bublitz and Reinhard Merkel, *Crimes Against Minds: On Mental Manipulations, Harms and A Human Right to Mental Self-Determination*, 8 Crim. L. & Phil. 51, 58 (2014) (imagining a situation where "a low dose of Ghrelin, not hazardous to bodily or mental health, is added" by a restaurant to its orders to "increas[e] appetite").

describe as a right of "mental integrity,' wherein the state enters a psychological realm that should remain free of its control.[72] However, it is also likely that other types of what we might call "freedom of mind" rights, which shield our right to engage in certain mental activity and perhaps change it with the aid of technology, are also right to be free of state interference in aspects of mental life that aren't most naturally described as a social practice.

The same may also be true of the exercise of our perceptual powers. Some of the observation and recording we engage in may be described as part of a social practice. This is certainly true of the vibrant and ongoing exchange of video recordings and other images that now occurs over social media. Drawing on Robert Post's account of First Amendment coverage as applying to specific social practices that advance First Amendment values, Seth Kreimer states that the First Amendment protects "courses of action that are recognized by social practice as comprising media of expression"—and then adds that "[i]n the last two generations, emerging technology and social practice have made captured images part of our cultural and political discourse."[73] We can also describe a more solitary viewing of images this way: The activity a person engages in when they use Google Earth or another global mapping program to immerse themselves in a computer-generated facsimile of a distant location, or to give themselves the sense of moving with that environment, can also be described as social practices that are now a familiar part of many individuals' lives.

But that may *not* be true of all of the ways we find value in seeing with technology. Imagine I use a new brain-computer interface or augmented reality visor to sense events not in my field of vision, and that can visually represent to me measurements of the air quality in a particular location. If a state prevents me from seeing that information, it is not clear that it would be intervening in an existing social practice. But it is still arguably undermining my interest in perceiving the world around me in such a way that it is crossing a constitutional line. It is preventing me from learning about the world and, especially if the information it is preventing me

---

[72] *See* Thomas Douglas and Lisa Forsberg, *Three Rationales for a Legal Right to Mental Integrity*. In: Ligthart, S., van Toor, D., Kooijmans, T., Douglas, T., Meynen, G. (eds) *Neurolaw*. Palgrave Macmillan (2021).

[73] Seth F. Kreimer, *Pervasive Image Capture and the First Amendment: Memory, Discourse, and the Right to Record*, 159 U. Pa. L. Rev. 335, 372 (2011).

from learning is not the kind that should be the subject of any privacy or security-shielding protection, then the state's actions are still arguably crossing a First Amendment or other constitutional line it shouldn't be permitted to cross.

The last example highlights another possible limit of focusing on social practice. When a person engages in conduct with new or uncommon technologies—perhaps technologies that the person has created for their own use—then it may be quite likely that no social practice will have developed around the use of such technologies. While it is certainly the case that individuals are beginning to make more use of extended reality (XR) technologies such as AR and VR, their use is still unfamiliar to most individuals (except perhaps in certain types of video games). While the use of video chat programs suddenly became pervasive after the Covid-19 pandemic, it is still rare for individuals to use technologies that are designed to give themselves telepresence in an immersive copy of a remote environment.

It is, of course, possible to characterize certain emerging technologies as new variations of familiar social practices. Playing a video game in virtual reality is arguably just a new way to play a video game. If so, it will be as staunchly protected by the First Amendment as is other kinds of video game play—which the Supreme Court made clear was shielded by First Amendment free speech protections in 2011, when it emphasized that "[l]ike the protected books, plays, and movies that preceded them, video games communicate ideas—and even social messages—through many familiar literary devices (such as characters, dialogue, plot, and music) and through features distinctive to the medium (such as the player's interaction with the virtual world)" and this "suffices to confer First Amendment protection."[74]

However, the question of whether to place an emerging technology within the coverage of freedom of thought, or the coverage of a freedom-of-thought-based right to see with technology, may not always be one that courts can answer by relying on a particular social practice. In some cases, they may have to instead reason from first principles regarding what values the right to freedom of thought or perception is designed to protect: To the extent that such freedom is understood as enabling individuals to have a certain measure of personal autonomy, they may have to describe

---

[74] Brown v. Entm't Merchants Ass'n, 564 U.S. 786, 790 (2011).

the nature of such autonomy—and draw conclusions about what kind of insulation we need against state interference with our thought and perception to safeguard it.

Or they might instead reason by analogy. They might ask if certain ways of augmenting our perception are analogous to kinds of perception that have been protected in the past. Consider an example. Seeing the world for ourselves often requires that we see with technology. Most of the events we learn about happen at times and places that prevent us from seeing or otherwise sensing them directly. That is why most of the history we know comes to us not through direct perception but through the accounts of historians, journalists, or other individuals. As philosophers describe this process, we gain knowledge of past events in the human or physical world—or sometimes, inaccurate accounts of it—through others' "testimony." "Testimony," writes Jonathan Adler, occurs when a person "makes an '[a]ssertion'" that puts "forth a proposition that the speaker represents as true."[75] As Michael Pardo has written, "A moment's reflection reveals that much of our knowledge is based on the testimony of others rather than on firsthand observations."[76]

That is why, in modern democracies, journalists must often act, as the Supreme Court has said, as the "eyes and ears of the public."[77] To make educated decisions in exercising their vote or otherwise exercising popular sovereignty, people need to know about numerous facts that underlie policy decisions—about wars, public safety, environmental protection, and many other matters of public importance. They generally rely on others' reports—and the testimony in these reports—to gather this information. When such reports come from the press, they have been staunchly shielded by the First Amendment's protection for freedom of speech. In fact, as Justice Potter Stewart has argued, such protection is more naturally rooted in the First Amendment's protection for "the freedom of the press."[78]

In the modern world, however, there are alternatives to relying on the press—and the verbal reports it provides about current events—to learn

---

[75] Jonathan Adler, *Epistemological Problems of Testimony*, Stan. Encyclopedia Phil., https://plato.stanford.edu/entries/testimony-episprob/.

[76] Michael S. Pardo, *Testimony*, 82 Tul. L. Rev. 119, 132 (2007).

[77] Houchins v. KQED, 438 U.S. 1, 8 (1978).

[78] Potter Stewart, *Or Of the Press*, 26 Hastings L.J. 631, 633 (1974–75).

about events in the world. Especially in a world where video technology is widely available and pervasive, audiences don't want to make do only with others' testimony. They instead want a video and/or audio recording that allows them to see or hear an event for themselves even when the event occurred far away. The rise of modern camera technology—of recordings and live video feeds—often allows people to insist that they be *shown* what occurred and not merely told about it. That is why many reports by the institutional press, whether in online newspapers or television shows, increasingly provide video footage of the events they report on. It is why, for example, many policymakers have responded to claims of police misconduct by demanding that police wear body cameras that produce a record of what occurred—so that those who assess whether there has been misconduct can view a visual record of what occurred. It is also why, when individuals make a claim about a matter of public importance on social media, they increasingly supplement their verbal reports about what has occurred with an audiovisual recording or photograph. And, as noted in various parts of the book, the First Amendment protects them when they do so. It shields this extension of seeing.

The question with which I ended the last chapter—and asked again near the beginning of this one—is whether this First Amendment protection should be *extended further*. Should the First Amendment protect individuals who receive remote visual data not only from others with whom they are communicating but also from a machine (such as a drone) under their control, or perhaps from some other means of engaging in a mechanically enhanced form of solitary information gathering? Barry McDonald has suggested that the press's traditional role as "the eyes and ears of the public" should, in the twenty-first century, be extended to other kinds of organizations or institutions that have the mission of gathering information on matters of public concern. The protection for information gathering extended to the press should, he argues, cover those who "society recognizes as performing legitimate and valuable information gathering and dissemination functions today (whether news media or not)."[79] This category might include academic researchers or think tanks.[80] It might also, I have suggested in the past, include those

---

[79] *See* McDonald, *The First Amendment and the Free Flow of Information, supra* note 10.

[80] *Id.*

who use technology to gather information, including visual data, such as Google and other creators of massive virtual maps.[81]

However, there is reason, I argue here, to extend the analogy to press rights even further. It is not just the intermediaries, such as mapmakers, who should be able to invoke a modernized freedom-of-press right to gather information of public concern. It is also the *individuals* who wish to use these intermediaries, or other technologies that augment perception, to engage in intellectual exploration themselves. In fact, they should be able to use this power not only to learn about "matters of public concern," but also to learn about matters of purely individual interest that are valuable to developing and revising their thinking process.

This analysis points us in a direction that we might use to formulate an alternative justification for technologies of enhanced seeing—one that goes beyond the free speech-based justification discussed in the previous chapter. Some of these justifications adhere closely to, and borrow heavily from, aspects of First Amendment doctrine courts have developed to find limits for the scope of free speech protection—or protection for information-seeking—where such limits are necessary. But as helpful as these ideas are, they too seem insufficient. Protection against illegitimate purposes doesn't answer the possibility government will often constrain seeing with plausible justifications that aren't focused solely on preventing us from gaining general knowledge about the world. Social practices that have value in exploring our environment might provide some guidelines for a right to see—but a future doctrine or freedom of thought or perception won't cover intrusions into those activities that are devastating for our autonomy without attacking particular social practices.

References

Articles and Books

Jonathan Adler, Epistemological Problems of Testimony, STAN. ENCYCLOPEDIA PHIL., https://plato.stanford.edu/entries/testimony-episprob/.
Anita L. Allen, *Dredging Up the Past: Lifelogging, Memory and Surveillance*, 75 U. Chi. L. Rev. 47 (2008).
Jane Bambauer, *Is Data Speech?* 66 Stan. L. Rev. 57 (2014).

---

[81] *See* Blitz, *The Right to Map, supra* note 34, at 123.

Jane R. Bambauer, *The Empirical First Amendment*, 78 Ohio St. L.J. 947 (2017).
Marc Jonathan Blitz, *The Freedom of 3D Thought: The First Amendment in Virtual Reality*, 30 Cardozo L. Rev. 1141 (2008).
Marc Jonathan Blitz, *Freedom of Thought for the Extended Mind: Cognitive Enhancement and the Constitution*, 2010 Wis. L. Rev. 1049 (2010).
Marc Jonathan Blitz, *The Right to Map (and Avoid Being Mapped): Reconceiving First Amendment Protection for Information-Gathering in the Age of Google Earth*, 14 Colum. Sci. & Tech. L. Rev. 115 (2013).
Marc Jonathan Blitz, *A Constitutional Right to Thought Enhancing Technology*, in ed. Veljko Dubljevic and Fabric Jotterand, *Cognitive Enhancement: Ethical and Policy Perspectives in International Perspective* (Oxford University Press 2016).
Marc Jonathan Blitz, *Free Speech, Occupational Speech, and Psychotherapy*, 44 Hofstra L. Rev. 681 (2016).
Jan Christoph Bublitz and Reinhard Merkel, *Crimes Against Minds: On Mental Manipulations, Harms and A Human Right to Mental Self-Determination*, 8 Crim. L. & Phil. 51 (2014).
Thomas Douglas and Lisa Forsberg, *Three Rationales for a Legal Right to Mental Integrity*. In: Ligthart, S., van Toor, D., Kooijmans, T., Douglas, T., Meynen, G. (eds) *Neurolaw*. Palgrave Studies in Law, Neuroscience, and Human Behavior. Palgrave Macmillan (2021).
David H. Freedman, *Why Scientific Studies Are so Often Wrong: The Streetlight Effect*, Discover Magazine, Dec. 9, 2010 [updated May 20, 2025], at https://www.discovermagazine.com/the-sciences/why-scientific-studies-are-so-often-wrong-the-streetlight-effect.
Seth F. Kreimer, *Pervasive Image Capture and the First Amendment: Memory, Discourse, and the Right to Record*, 159 U. Pa. L. Rev. 335 (2011).
Simon McCarthy-Jones, *Freethinking: Protecting the Freedom of Thought Amidst the New Battle for the Mind* 55 (2023).
Barry P. McDonald, *The First Amendment and the Free Flow of Information: Towards a Realistic Right to Gather Information in the Information Age*, 65 Ohio State L.J. 249 (2004).
Michael S. Pardo, *Testimony*, 82 Tul. L. Rev. 119 (2007).
Robert Post, *Recuperating First Amendment Doctrine*, 47 Stan. L. Rev. 1249 (1995).
Natalie Ram, *Science as Speech*, 102 Iowa L. Rev. 1187 (2017).
Dana Remus Irwin, *Freedom of Thought: The First Amendment and the Scientific Method*, 2005 Wis. L. Rev. 1479.

Cases

Brown v. Entm't Merchants Ass'n, 564 U.S. 786 (2011).
Chaplinsky v. New Hampshire, 315 U.S. 568 (1942).
Doe v. City of Lafayette, Ind., 377 F.3d 757 (7th Cir. 2004).
Houchins v. KQED, 438 U.S. 1 (1978).
Hurley v. Irish-Am. Gay, Lesbian, Bisexual Group of Boston, 515 U.S. 557 (1995).
National Press Photographers Ass'n v. McCraw, 90 F.4th 770 (5th Cir. 2024).
R.A.V. v. City of St. Paul, 505 U.S. 377 (1992).
Richmond Newspapers v. Virginia, 448 U.S. 555, 576 (1980).
Sable Commc'ns of California, Inc. v. F.C.C., 492 U.S. 115 (1989).
Spence v. Washington, 418 U.S. 405 (1974).
Stanley v. Georgia, 394 U.S. 557 (1969).
United States v. O'Brien, 367 U.S. 376 (1968).
United States v. Playboy Entm't Grp., Inc., 529 U.S. 803 (2000).
United States v. Stevens, 559 U.S. 460 (2010).
Va. State Bd. of Pharmacy v. Va. Citizens Consumer Counsel, 425 U.S. 748 (1976).
Virginia v. Black, 538 U.S. 343 (2003).
Zemel v. Rusk, 381 U.S. 1 (1965).

CHAPTER 4

The Right to Natural and Extended Vision (and Bodily and Mental Integrity)

Abstract The previous chapters have focused on the First Amendment's protection for freedom of speech (and unexpressed thought and knowledge creation) as possible bases for a right to use our perception to observe the world—and to augment that perception. This chapter takes a different approach. It begins with a right that different provisions of the Constitution (the Fourth, Fifth, and Fourteenth Amendments) give us: a right (or rights) to bodily freedom and personal integrity. It then points out that these provisions should not only shield our bodies and persons but more broadly shield the prosthetics we use to replace lost body functions, as well as other technologies we integrate into our bodies (and perhaps, make a part of our minds). The challenge arises when we explore extending such protection to technologies that aren't restorative and integrated into us—but rather are at a further remove and enhance our vision. These extensions may seem less integral to our autonomy and may also threaten other interests in the world.

Keywords Due process · Autonomy · Integrity · Personal liberty · Extended mind · Enhancement · Harm · Brain-computer interface · Fifth amendment · Fourteenth amendment · Surveillance · Prosthetic · Hacking · Supreme Court · Constitutional rights

© The Author(s), under exclusive license to Springer Nature Switzerland AG 2025
M. J. Blitz, *The Right to See with Technology*, Palgrave Studies in Law, Neuroscience, and Human Behavior,
https://doi.org/10.1007/978-3-031-89533-3_4

Do we have a constitutional right to perceive the world around us—directly or by using technology? If so, what is the source of such a right? In the previous two chapters, I looked at two answers to this question, both of which seemed problematic. Courts have grounded a certain kind of enhanced vision—namely, recording our environment—in the right that the First Amendment gives us to speak: In the twenty-first century, the recordings we share, on web pages, social media posts, and other computer applications, are a common medium of communication. So creating the recordings is "speech creation."[1] This answer, however, leaves unanswered the question of how the Constitution applies to many other forms of augmented perception that help us see, but don't (primarily) serve the purpose of communicating what we see. They take account of our interest in conveying our perspective on the world to others, but not our equally strong interest in silently using perception to observe and learn about the world around us, or to shape and reshape our memory and cognition more generally. Recording itself sometimes is most valuable to use for purposes other than communication.

For that reason, we might propose a broader foundation for a right to see: It is a part and parcel of the right to obtain information, to learn about the world through the exercise of a broader conception of the First Amendment that protects not just free speech but intellectual liberty. This response, however, seems to make the right *too* broad. It is not only the perception of our immediate surroundings that adds to our memories and our more general stock of information about the world: It is practically everything we might do. Spelunking in a cave or skydiving. Using a certain kind of illicit drug. All of these allow us to have experiences that we otherwise couldn't have, form memories we otherwise wouldn't have, and learn things we otherwise couldn't learn. That cannot mean that all of these count as exercises of a right to gather information protected by the First Amendment. If so, it is hard to understand what exercises of our perception are not protected by the First Amendment since, as the Supreme Court noted in Zemel v. Rusk, restriction of any activity could be characterized as resulting in "decreased data flow."[2] *Any* restriction

---

[1] See Irizarry v. Yehia, 38 F.4th 1282, 1289 (10th Cir. 2022) ("videorecording is 'unambiguously' speech-creation, not mere conduct" and is thus protected under the First Amendment"); W. Watersheds Project v. Michael, 869 F.3d 1189, 1196 (10th Cir. 2017) (recording "fit[s] comfortably in the [category of] speech-creation").

[2] *Zemel v. Rusk*, 381 U.S. 1, 15 (1965).

stops us from performing some act that would leave us with distinctive memories and knowledge of what we encounter. There are certain principled limits that scholars have proposed placing on a right to receive information. They have suggested, for example, that knowledge creation can be protected from government attacks targeting it specifically rather than incidentally limiting it while addressing another type of government interest.

However, many hypothetical restrictions or distortions of our capacity to perceive our environment seem intuitively to violate our fundamental freedoms. As I noted in earlier writing, if the government banned our use of telescopes to view the skies, it would seem to be violating a key commitment of a free society.[3] The same would be true if it barred us from making observations of the flora or fauna in a particular environment, whether with our unaided visions or the help of binoculars. Or if it forbade us from using virtual reality technology to give us telepresence in environments distant in space or time to learn about it in ways that don't violate others' privacy interests. If the wrong or dubious nature of these restrictions should make them constitutionally and not just morally suspect, what account can one give for that conclusion if neither a speech-based nor a general information-gathering-based right can provide a foundation for it?

This chapter will explore doing so by beginning closer to the specific conduct being restricted in the examples I've just used: The use of one's biological vision in certain ways, either by itself or enhanced to some degree with technology: Eyeglasses, binoculars, telescopes, or virtual reality and other computer-mediated kinds of seeing. It will also suggest that in thinking about our right to use and augment our natural vision in this way, recent scholarly analyses on the role technology—and particularly, technology enabled by modern neuroscience—plays in our freedom of thought or "cognitive liberty" provide both a model for a right to see and another foundation for it: A right to see with technology can be conceived along the same lines as a right to think with technology, and there is a case to be made that the former is a component of the latter.

---

[3] *See* Marc Jonathan Blitz, *The Freedom of 3D Thought: The First Amendment in Virtual Reality*, 30 Cardozo L Rev. 1141, 1154 (2008); Marc Jonathan Blitz, *The Right to Map (and Avoid Being Mapped): Reconceiving First Amendment Protection for Information-Gathering in the Age of Google Earth*, 14 Colum. Sci. & Tech. L. Rev. 115 (2013).

The chapter will do so by exploring two different ways to understand seeing or sensing: (1) The way it is realized by certain biological mechanisms—to which human beings have added an array of technologies that extend our natural vision in certain ways—and (2) seeing and other perceptions as a certain kind of function that might be understood as separable, in some respects or circumstances, from the biology that makes it possible—and protected by a right to see that follows this function, and shields it from government restriction or control, even when it occurs outside of our body. As Chapter 5 explains more fully, how a system of rights protects the function of seeing may in part be a function of social practices or conventions, which may in turn be affected by whether certain technologies are—as the Supreme Court has put it in its Fourth Amendment case law—"in general public use."[4] In the course of setting out this analysis, I will also draw analogies to similar arguments that have been made about a right to freedom of thought or cognitive liberty.

THE RIGHT TO SEE AND BODILY AUTONOMY AND INTEGRITY

In debates about and analyses of a right to "freedom of thought" or "cognitive liberty," arguments are sometimes made that we don't need a new right to the mental freedom covered by such concepts because our right to think is *already* protected by two more familiar rights.[5] It is protected from attacks that state officials make against the thought we express in words: When we express our beliefs or other ideas in language, the state is barred by punishing us for forming or having those ideas by the protection the First Amendment gives to "freedom of speech." Once thoughts are no longer hidden inside of our minds but rather take a form—in words or other expression—where the state can identify them as ideas it opposes, then these thoughts are no longer shielded from punishment

---

[4] *Kyllo v. United States*, 533 U.S. 27, 34 (2001) (finding that use of thermal imaging or other technology that allows viewing the home's interior would be a "search" constrained by the Fourth Amendment—but suggesting that would not necessarily be true where such technology was already "in general public use").

[5] I have previously reviewed these arguments in Marc Jonathan Blitz, *Freedom of Thought for the Extended Mind: Cognitive Enhancement and the Constitution*, 2010 Wis. L. Rev. 1049, 1090–1111 (2010).

by nature (since they are now knowable by officials) but they are shielded by protections against censorship.

There are, to be sure, other ways the state may attack our thinking: If it tries to control our thought by manipulating our brain activity, then it need not wait for our thoughts to be expressed in words to target them. Rather than waiting for us to *externalize* our thought in language, it can attack them—and the psychological processes that generate them—even when they remain *internal* to us. In this circumstance, however, the state will face another constitutional barrier: The shield that the Constitution gives us against intrusion into our *bodies*. The brain, of course, is a part of the body. So if the state tries to force us to take a psychotropic medication to change the way our brain functions, it must overcome the legal force field American law generates to protect bodily integrity and autonomy.

This shield is not solely a constitutional one. The Supreme Court wrote in 1891, in the case of *Union Pacific Railway Company v. Botsford*: "No right is held more sacred, or is more carefully guarded by the common law, than the right of every individual to the possession and control of his own person, free from all restraint or interference of others unless by clear and unquestionable authority of law."[6] One may invoke this right to sue when physical integrity is violated, or one's physical liberty restrained, by other private parties. But one can also invoke a constitutional version of this right to oppose intrusion in one's body, or interference in one's physical liberty, by the state. The Supreme Court has quoted *Botsford* in emphasizing that the Fourth Amendment's protection against unreasonable search or seizure prevents the state from detaining or searching a person without strong justification of the kind demanded by that constitutional provision. As it noted in *Terry v. Ohio*, a case where a person was subjected by a law enforcement officer to a "pat down" search to see if he was armed, people have an "inestimable right of personal security" and they retain this right outside of their homes.[7] That, it said, is made clear by *Botsford's* emphasis on "the right of every individual to the possession and control of his own person," as well as by other Supreme

---

[6] Union Pac. R. Co. v. Botsford, 141 U.S. 250, 251 (1891). The Court quoted a statement by Judge Thomas Cooley, in an 1888 treatise on tort law, stating that "[t]he right to one's person may be said to be a right of complete immunity; to be let alone". *A Treatise on the Law of Torts* 29 (2d ed. 1888).

[7] *Terry v. Ohio*, 392 U.S. 1, 8–9 (1968).

Court precedent.[8] Courts have also repeated *Botsford's* language in cases involving compelled blood tests for drivers suspected to be drunk.[9]

It has likewise stressed the right to bodily integrity and autonomy in cases on the liberty protected by the due process clauses of the Fifth and Fourteenth Amendments. It made this clear in *Rochin v. California*, for example, a 1952 case in which police—in an effort to make a criminal suspect vomit out drug capsules he had just swallowed—"forced an emetic solution through a tube into Rochin's stomach against his will."[10] The Court emphasized that, because the due process clause of the Fourteenth Amendment safeguards certain "personal immunities" that may not be violated by the state, courts have to understand what liberty is entailed by those immunities and guard its boundaries.[11] In this case, the Court said, the state had clearly violated Rochin's constitutional liberty: This was "conduct that shocks the conscience." It involved "[i]llegally breaking into the privacy of the petitioner," then forcibly "open[ing] his mouth and remove what was there, the forcible extraction of his stomach's contents."[12] Although officials can use certain methods to collect evidence of crimes they cannot constitutionally do so in a way that violates bodily integrity in this way.

The Court has also protected our bodily integrity in other Fourteenth Amendment "due process" cases, including those that have a closer relationship to freedom of thought concerns. It has held that the state cannot *force* prisoners or those committed to psychiatric institutions to take antipsychotic medications unless it can make a showing that such forcible medication of an individual is necessary to protect their safety, that of other people, or to satisfy some other weighty state interest. In *Washington v. Harper*, in 1990, for example, the Court stressed that "[t]he forcible injection of medication into a nonconsenting person's body represents a substantial interference with that person's liberty" even though "[t]he purpose of the drugs is to alter the chemical balance in a

---

[8] *Id.* at 9.

[9] *See, e.g.*, State v. Villarreal, 476 S.W.3d 45, 49 (Tex. App.2014).

[10] Rochin v. California, 342 U.S. 165, 166 (1952).

[11] *Id.* at 169–173.

[12] *Id.* at 209–210.

patient's brain, leading to changes, intended to be beneficial, in his or her cognitive processes."[13]

This doesn't mean that searches or other measures that restrain physical liberty or intrude into bodily integrity can *never* take place. The Supreme Court ultimately permitted the warrantless pat down in *Terry* and allowed police to conduct blood tests on suspected drunk drivers if they can obtain a warrant (or an "exigent circumstances" exception to the warrant requirement applies).[14] It has even allowed compelled *random* drug testing of employees or high school students to prevent drug use where that would create a threat to safety.[15] Even the forcible administration of psychoactive drugs in *Harper* was found by the Court to be justified in the end: It relied on a judgment by medical decision-makers that such treatment was necessary and allowed the defendant to contest the need for drugs at a hearing.[16] However, even if our right to bodily integrity is not an absolute right it *is* a meaningful barrier against state interference.

I have briefly described these two existing safeguards for our *thinking*—(1) the protection the free speech clause gives us against punishment of the *thoughts we express* and (2) the autonomy-based protection of the *biological processes that produce the thinking*—because they suggest an initial possible response to the conundrum that the previous chapter finished with, and that this chapter began by summarizing. If we turn our focus from a right to think to a right to see, we may find that this double protection can also provide guidance in developing that right. If, as the chapter said, an attempt to ground a right to see in the First Amendment's protection for speech or information gathering is problematic in certain respects, perhaps we can—as in the protection of our thoughts—find an alternative in protection for the *biology* that underlies seeing, and more generally for our right to freedom from state interference in a realm of personal autonomy and privacy.

---

[13] Washington v. Harper, 494 U.S. 210, 229 (1990).

[14] *See* Terry v. Ohio, 392 U.S. 1, 23–25 (1968); Birchfield v. North Dakota, 579 U.S. 438, 474–476 (2016).

[15] *See* Vernonia Sch. Dist. 47 J v. Acton, 515 U.S. 646 (1995), Skinner v. Ry. Labor Executives' Ass'n, 489 U.S. 602, 628 (1989), Pottowatomie County v. Earls, 536 U.S. 822, 833 (2002).

[16] *Washington v. Harper*, 494 U.S. 210, 231–236 (1990).

Even where the government can argue persuasively that its regulation of a certain type of perception (natural or augmented) is not a restriction of speech, and does not restrict information gathering any more than many other unquestionably legitimate restrictions of conduct still, we might argue, it impermissibly intrudes into our own control over own body, or of deeply personal exercises of our autonomy which the state generally has no justifications to interfere with, when it tries to restrict where we turn our gaze. Individuals' eyes are designed to receive light from the environment around them. Their nervous system processes that input and builds a picture of the outside world from it. As a general matter, the state has no constitutionally permissible ground for controlling the functioning of that biologically and psychologically essential system for perceiving the world any more than it has grounds for intervening in the way our brains produce thoughts. It is only when the state has extraordinary justification to enter that realm that it may, perhaps, exert some control over our perception.

This then provides a starting point for a somewhat different way of thinking about a right to see and perhaps, to see with technology than the one I have explored in the preceding chapters. The core of that right is not a right to communicate about what we see. Or, if that is a core element of it, it is not the *only* core one finds there. Just as crucial is a right of personal privacy—a right to be let alone. That right gives us control over our own bodies and perhaps establishes a certain space or "buffer zone" around our bodies that prevents the state, as well as other external actors, from violating or posing a significant threat to our autonomy. It prevents the state from somehow inserting itself into our visual processes and other sensory systems in some way equivalent to forcibly inserting itself into a person's brain functioning with compelled administration of psychotropic drugs. And it prevents it also from commanding us to change the way we use these sensory systems, and perhaps from forbidding us from at least some technological alterations of it—for example, by barring us from wearing glasses to improve our vision or pointing a set of binoculars or a telescope at a part of the natural world.

The last proposal about perception will likely elicit an objection: There may be cause to worry about new "neurointerventions" that manipulate our brain activity. But there is little cause, some might argue, to worry that the target of such intervention is generally our natural vision or visual processing system. Why would we need "the right to be let alone" to ensure our natural biological vision remains undisturbed by government officials when there is little evidence that they have the means or desire to disturb it? Conceived that way, the right to see—one might argue—provides us with a fortified constitutional wall against a non-existent threat. Why, then, should courts take the trouble to build such a wall and why should legal thinkers invest intellectual energy in deliberating about how it should work?

Regulations that target our perception, in the United States and other jurisdictions, have focused on use of cameras—in cell phones, on drones, or worn on our person—to make recordings of others. It is that kind of recording that the government has sometimes sought to restrict, when it might capture evidence of government activity that officials have argued they can keep from being recorded. And it is that kind of recording that many others claim needs to be understood as a threat to privacy—and regulated for that reason—for example, when drones can be potentially used to capture recordings of a neighbor's backyard (or someone else's property).[17] So the type of augmented seeing that has received an explicit constitutional shield from courts in some cases—and that needs that protection—is the kind of recording that was the focus of Chapters 2 and 3 and can be used to share information—to communicate.

This objection, however, has too limited an understanding of the way emerging technologies are augmenting our senses and may become more common and powerful—and the interest that underlies such augmentation (and may even underlie many contemporary image capture). Consider enhancements of vision that occur in brain-computer interface devices, or with wearable technology that lets us see things normally invisible to us. Or lifelogs that we create to supplement our memory.[18] Or uses of extended reality technology to alter the source of our sensations—effectively placing us in distant settings.[19] The fast-paced improvement

---

[17] See Margot E. Kaminski, *Drone Federalism: Civilian Drones and the Things They Carry*, 4 Cal. L. Rev. Circuit 57, 72 (2013) (noting that because of their relatively low cost and hovering abilities, drones give rise to a specter of pervasive surveillance).

[18] *See* Allen Coin and Veljko Dubljević, *An Introduction to Policy, Identity, and Neurotechnology: The Neuroethics of Brain–Computer Interfaces*, in ed., Allen Coin and Veljko Dubljević, *Policy, Identity, and Neurotechnology: The Neuroethics of Brain–Computer Interfaces* (Springer 2023).

[19] *See* U.S. Government Accountability Office, Science & Tech Spotlight: Extended Reality Technologies, Jan. 26, 2022, at https://www.gao.gov/products/gao-22-105541; *See* Telepresence: Hearing Before the Subcomm. on Science, Technology and Space, 105th

of artificial intelligence and computer technology may increase the likelihood that these technologies will become more prominent parts of our lives. Think about smartphones and their capacities—the capacities they've created to be in contact with others or be contacted most of our waking hours, the always-on internet connection that lets us find information that may have once required a trip (or many trips) to a library or libraries, the GPS technologies that allow individuals to navigate their environment. These have, in little more than a decade, become essential for many individuals to meet the demands of work and family life. As the Supreme Court said in one case on searches of cell phones, "they are now such a pervasive and insistent part of daily life that the proverbial visitor from Mars might conclude they were an important feature of human anatomy."[20] This is already arguably true of the additional capacity smartphones give us to capture pictures or videos anywhere we go, to instantly review or share them, and to store them inside of our phone (or on "the cloud" of computers that exist for such purposes). It might also become true of some of the other forms of augmented seeing discussed in this book. If and when that becomes true, a speech rights-only foundation for a right to perceive will, this book has argued, be inadequate. So it is important to understand alternatives—one of which might ground a right to see with technology (which may well need protection from government) in a more basic right to see with the natural powers generated by the biology of each person (which are generally protected by what the right to "inviolability of the person" and the buffer zone around it). So, in the sections below, the chapter will begin with the protection the law gives to our bodies and personal space and work outwards from that starting point to technological transformations and additions to our natural capacities.

Cong. 14 (1998) (statement of S. Kicha Ganapathy, Member, Technical Staff, Multimedia Communications Research Laboratory, Bell Laboratories).

[20] Riley v. California, 573 U.S. 373, 403 (2014).

Protection for Our Bodies—And Protection for Seeing That Comes with It

Whatever else our personal autonomy includes, it includes, as the above case law emphasizes, control over our own bodies and minds. That doesn't mean the law will protect all decisions we make and everything we choose to do with, or to change, our bodies and minds. (The Supreme Court has recently stressed that constitutional protection for autonomy cannot be understood so broadly that it shields "prostitution" and "illicit drug use" from the restrictions to which they have long been subject in the United States.)[21] But it does mean that we are protected against certain kinds of interference in our person: As noted above, the state cannot—without overcoming constitutional hurdles—subject us to compelled medical treatment or physically restrain us.[22] Moreover, while the autonomy we receive under the Constitution encompasses, a minimum, what we do with our bodies *and minds*, I will focus initially on our physical freedom. Outside the realm of free speech and religion, constitutional law has had little to say about any independent protection for freedom of thought. Modern scholarship on freedom of thought has had more to say about that liberty. But it is helpful to start with a narrower focus—on our bodily autonomy and protection of our persons against physical interference.

Such protection rules out at least certain kinds of interference in the way we exercise our powers of perception. Others cannot forcibly blindfold us without committing the tort of battery. Nor could the state do so, or force us to undergo surgery affecting our perceptual power, without likely violating Fourteenth Amendment rights protecting our bodily integrity.

This suggests one way that seeing *with technology* can be protected by the Constitution even when it has little to do with speech or other expression. When such technology is so integrated into our body that it becomes a part of it—or we become so reliant upon it in exercising our bodily freedom that it becomes a necessary condition of that freedom—then a state's interference with that technology seems intuitively to be as

---

[21] Dobbs v. Jackson Women's Health Org., 597 U.S. 215, 218 (2022).
[22] Cruzan v. Director of Missouri Department of Health, 497 U.S. 261 (1990).

much interference in our person as is any restraint of, control over, or damage to, the functioning of our natural organs.

This seems clear in cases where we face interference in more familiar technologies that allow our bodies to function. Cybersecurity and medical experts as well as television writers have already imagined the frightening possibility that malicious computer hackers might gain access to and disable pacemakers surgically implanted in hearts.[23] In people who need them, such pacemakers partially replace the natural electrical activity of the heart to ensure it doesn't beat too slowly. When a computer hacker exerts control over such an artificial device to end or threaten a person's life, or create a serious risk to their survival, they are quite clearly violating the bodily integrity of the person in which the pacemaker is implanted. That they are aiming their attack at a machine implanted within the body rather than natural tissues makes little difference. The damage they are doing is just as serious as (if not identical to) the damage they would do if they found a way to alter the rhythm of a beating heart that lacked a pacemaker. Whether by attacking a pacemaker or the heart's tissues, they are causing bodily harm that violates a person's right against it—and giving themselves control over a person's functioning they have no right to have. An outsider's damage to or control over a computerized prosthetic arm or leg would offend our "personal inviolability" for the same reason. By replacing a natural limb we've lost, the prosthetic arm or leg has become a part of our body—and must remain free of interference or control by others if we are to retain our bodily integrity and freedom.

Writing on brain-computer interface devices has made an analogous point about potential hacking into those devices. Tamara Bonaci, Ryan Calo, and Howard Jay Chizek, for example, have warned that BCI could allow hackers to control brain processes or insert spyware to gather information from a BCI.[24] A recent piece by J. Adam Carter imagines how a neurolink chip or other implant in our brains might violate our freedom of thought if we rely upon it to provide information that we integrate

---

[23] *See*, e.g., Subrat Das et al., *Cybersecurity: The Need for Data and Patient Safety with Cardiac Implantable Electronic Devices*, 18(3) Heart Rhythm (March 2021): 473–481.Homeland, Season 2, episode 10 (*"Broken Hearts"*; The Blacklist, Season 6, Episode 10 ("The Cryptobanker").

[24] Tamara Bonaci, Ryan Calo, and Howard Jay Chizeck, *App Stores for the Brain: Privacy & Security in Brain-Computer Interfaces*, Proceedings of the IEEE 2014 International Symposium on Ethics in Engineering, Science, and Technology (2014).

into our thought. It could do so, he says, either through "acquisition manipulation"—whereby our autonomy in forming beliefs is undermined by "having beliefs or desires 'implanted' in a clandestine fashion" or "eradication manipulation" wherein beliefs or desires are "'wiped' in a clandestine fashion."[25] Where the hacker or manipulator is a state official rather than a hacker who is a non-state actor, the infringement of personal liberty is no less significant. In fact, as terrible as hacking targeting an individual's body would be, liberty is often subject to even greater threat when the entity that is supposed to be protecting against such intrusion into our bodily integrity is instead the source of it—and seeks to exercise that kind of control over the citizenry as a whole.

The same point can be made about machines integrated into our bodies to replace lost vision function. Significant research is now being done on artificial eyes that could restore vision to those who have lost it or give vision to those who never had it. For example, Allen Coin and Veljko Dubljevic write, BCI can restore "perceptive sense" with devices such as "cochlear and visual cortical implants" and might also allow individuals to perform tasks previously beyond human capacities.[26] A recent prototype of a bionic eye works by "feeding signals" from a camera "directly to the brain."[27] This is currently meant to restore sight rather than improve it. But there is no reason the same technology might not one day give us new capacities, such as the ability to perceive infrared or ultraviolet light.[28]

It seems quite clear that the interference of another person—whether a private or state actor—in the operation of such an artificial eye would interfere with bodily integrity in the same way as interference with a pacemaker, a prosthetic limb, or a brain-computer interface that controls our muscle movement. Actions by state officials or individuals to control such

---

[25] *See* J. Adam Carter, *Varieties of (Extended) Thought Manipulation*, in ed. Marc Jonathan Blitz and Jan Christoph Bublitz, *The Law and Ethics of Freedom of Thought, Volume 1: Neuroscience, Autonomy, and Individual Rights* (Palgrave Macmillan 2021).

[26] Allen Coin and Veljko Dubljevic, *The Authenticity of Machine—Augmented Human Intelligence: Therapy, Enhancement, and the Extended Mind*. Neuroethics. 2021; 14: 283–290.

[27] *Id.*

[28] Russ Juskalian, *A New Implant for Blind People Jacks Directly into the Brain*, MIT Technology Review, Feb. 6, 2020, at https://www.technologyreview.com/2020/02/06/844908/a-new-implant-for-blind-people-jacks-directly-into-the-brain/. *See* Coin and Dubljevic *supra* note 26.

technology to manipulate what we see would clearly infringe our personal autonomy. We likely don't need to invoke a distinctive stand-alone right to see or perceive in order to identify and counter the wrongness of such actions: The same right to bodily integrity and autonomy that allows us to classify interference with artificial heart or limb components as impermissible, and counter it, would allow us to do the same for interference in an artificial eye, component of the eye, or for that matter, any other artificial sensing device we use to replace lost sensing capacity by implanting it into our body.

Visual Processing, Brain, and Environment

Before considering additional questions about when we might have a right to replace or supplement our natural visual systems, it is helpful to highlight a few aspects of the natural visual systems that such technology might replace. We do not, for purposes of this discussion, need to deeply explore the biology of our visual system. Just as we can discuss and consider claims about freedom of thought without anything close to a complete account of the brain processes that generate thought, so we can explore whether we have a right against interference in our perception without a thorough account of the biology that underlies visual perception or other kinds of sensation. However, in thinking about how we can enhance our powers of perception, whether and when we have a constitutional right to do so, and what the nature of that right would be, it is useful to highlight a few features of those biological processes. This is especially true if we want to understand how certain kinds of technologies of perception might be used to address certain challenges that an individual may face in perceiving their environment—whether these challenges arise from certain idiosyncratic features of their own visual processes or features that characterize human vision more generally.

First, in a world of cameras, movies, and display screens, we often tend to understand our natural technology of seeing by analogizing it to these more familiar—and in most respects, simpler—processes. Doing so carries the risk that we will understand vision or other senses as being more passive than they really are—as simply absorbing and transmitting a picture of the world around us to our consciousness rather than actively constructing it, in a complex process that transmits only some of the visual stimuli we receive to conscious awareness and that shapes the image it constructs for our practical needs. Writers who describe the biology of vision have emphasized that common analogies are misleading. As Margaret Livingstone observes, "It is commonly thought that the eye is like a camera, and that its function is to send a high-resolution image of the visual world to the brain" but this a "misperception:" The light the eye captures from the outside world is not merely captured and delivered but processed by the nervous system so that the information acquired through vision can be reflected in a "biologically useful [ ] way" that allows the organism to interact with the environment.[29]

This process begins with light entering our eyes. As it does so, the light passes through a number of structures that help control the passage of light on its way to the retina: The cornea, the iris and pupil, the lens, and the fluid-filled spaces that allow the eye to maintain its spherical form. The muscles and nerves in the iris control how much light passes through by controlling the size of the opening, the pupil, that lies at its center. The lens and cornea, as John Dowling and Joseph Dowling note, "are responsible for the focus of the eye; that is, they bring the rays of light emanating from objects at different distances to a sharp point on the photoreceptor cells" within the retina.[30]

The retina itself is where the processing of the visual signal begins. It consists of a thin, intricate sheet of cells at the back of the eye, where many different types of neurons begin to analyze it. The light first reaches photoceptor cells: Rods and cones, which get their names from the shapes of the categories of cells they refer to. There is a division of labor of sorts between different cells and different regions of the retina. Rods are more sensitive to light than are cones and detect information about the brightness—or amount of light—generated in a part of our visual field. Their sensitivity allows us to see in dim conditions.[31] Cones play a greater role in allowing for high-resolution vision and also allow us to see color. As Livingstone notes, "[m]ost of us have three different kinds of cones each of which contains a different kind of pigment and responds best over a different range of visible wavelengths." These cone types are often

---

[29] Margaret S. Livingstone, *Vision and Art* (Updated and Expanded Edition) 24 (Abrams 2014).

[30] John E. Dowling and Joseph L. Dowling, Jr., *Vision: How It Works and What Can Go Wrong* 49 (MIT Press 2016).

[31] Richard L. Gregory, *Eye and Brain: The Psychology of Seeing* 53 (Princeton Science Library Book 80) (noting that "rods function under low illumination, giving vision only of shades of grey.")

labeled based on the wavelength of light they receive as "red," "green," and "blue" cones and our nervous system senses colors by comparing the activation of different types of cones.[32] As Dowling and Dowling explain, this kind of differentiation allows our vision to be "remarkably adaptable." When we are in a dark environment, "[t]he retinal rod photoreceptors become exquisitely sensitive, capable of responding to just a single photon. When the eye is exposed to light, the cone photoreceptors come into play, and we see details and color."[33] There is also a spatial specialization. A small depression at the center of the retina—the fovea—is packed with cones and is the area where our vision is the sharpest. The visual acuity made possible by this region is crucial for activities such as reading and driving a car.[34] By contrast, areas outside of the fovea allow us to see with peripheral vision—which does not have the same resolution as foveal vision does but allows us to see other parts of our environment that may be important for us. Before the visual information registered by the photoreceptors leaves the retina, it is sent, via bipolar and amacrine cells, to retinal ganglion cells the fibers or "axons" of which form the optic nerve that carries visual signals to the brain, primarily to a region of the thalamus called the lateral geniculate nucleus which then relays information to the "visual cortex" of the brain. Even before this visual information leaves the eye, the cells receiving information from the photoreceptors begin processing it,to determine the illumination and colors of the objects we're viewing, the contrast between light and dark portions of what we see, shapes, movements, and other features of what is in our visual field.

The photoreceptors' response to light is only the beginning of a much more complex process, wherein visual signals travel first from photoreceptors to retinal ganglion cells (by way of other cells, such as bipolar cells) and then to the brain along the wire-like filaments or "axons" that extend from bodies of ganglion cells along the optic nerve to structures in the brain. A small number of these axons link the retina to a structure in the midbrain called the superior colliculus that appears to play an important role in directing the gaze of the eyes. But most of these

---

[32] Livingstone, *supra* note 29, at 28, 38.

[33] Dowling and Dowling, *supra* note 30 at 29.

[34] *Id.* at 37.

nerve strands connect to a structure in the thalamus, the lateral geniculate nucleus, which in turn connects to the part of the cerebral cortex that is focused on processing vision (the visual cortex). The visual cortex is in turn subdivided into areas within the brain that play specialized roles in visual processing—with distinctive sets of cells processing shapes, movements, and colors of objects. Neuroscientists have described different areas of the cortex involved in visual processing: V1, V2, V3, V4, and the medial-temporal (MT) area.[35] Other parts of the brain also play a role in our experience of vision. For example, our visual memory of the past draws on brain processes apart from those that let us see what is in front of us in the present. Moreover, perceiving the world successfully generally requires not only detecting the shapes, movements, colors, and other visual features of our environment (or remembering them from a past experience) but also weaving them together with other stimuli—the sounds we hear and olfactory and tactile stimuli we receive—and with information that allows us to recognize a part of the visual scene as an item of food we can safely eat or a dangerous object to avoid.

Because each step of this complex biological process plays an important role in our capacity to see, recognize, and respond to our environment, damage to any part of it can impair our vision. Individuals who lack any of the three types of cone cells, for example, will have some form of color blindness. Inherited or age-related conditions that interfere with the functioning of photoreceptor cells in other ways can lead to severe loss of vision. Retinitis pigmentosa (RP), for example, is an inherited disease that, as Dowling and Dowling explain, involves the gradual deterioration of photoreceptor cells—first rod cells, with a resulting loss of night and peripheral vision—and then cone cells, causing deterioration of acute and color vision. Macular degeneration most frequently occurs in aging individuals and entails damage in the macula, the part of the eye that contains the fovea and allows for central acute vision.[36]

Damage to parts of the brain involved in visual processing can result in neurological problems. These can include agnosia (whereby the brain detects but cannot recognize persons, objects, shapes, or other visual information), akinetopsia (or "motion blindness"), or "neglect" or an

---

[35] *See* Christoph Koch, *The Quest for Consciousness: A Neurobiological Approach* 57-63 105-152 (Roberts & Co, 2004); Dowling and Dowling, *supra* note 30, at 32.

[36] Dowling and Dowling, *supra* note 30, at 85-92, 124.

inability to perceive and react to what the eyes sense in one part of the visual field. Given the central role that vision and other forms of sensation play in our lives, any of these defects in our perception can have devastating effects.[37]

Conceivably, technology could thus play a role in replacing or augmenting any part of this biological process. As Lauren Ayton and her co-authors note in an overview of retinal implants, attempts to artificially restore vision go back centuries: An eighteenth-century French physician, Charles LeRoy, attempted to cure blindness by delivering electrical impulses to the nervous system of a blind subject, but only caused the subject to perceive flashes of light. Progress was made over the course of the twentieth century in exploring how vision might be restored with devices that deliver electrical signals directly to parts of the nervous system.[38] In recent decades, there has been significant work on developing artificial retinas that can in some way replace photoreceptor cells that have failed in diseases such as retinitis pigmentosa. In 2013, for example, the Food and Drug Administration of the United States (FDA) approved the Argus II device. This device bypasses the no-longer functioning photoreceptor cells by delivering visual signals to retinal ganglion cells from a surgically implanted array of microelectrodes that in turn receives visual information from a camera mounted on glasses and a connected visual processing unit. Other artificial retina prototypes are implanted at different locations in the retina.[39] Engineers have also developed brain-computer interface (BCI) devices that aim to generate some useful visual information in blind individuals by linking a camera directly to the visual cortex.[40] One can imagine other BCI technology in the future being designed to correct problems in vision that arise from injuries or diseases in the visual cortex.

---

[37] *See* Adam Zeman, *Consciousness: A User's Guide*, 208-223, 377 (Yale University Press 2002).

[38] Lauren N. Ayton, et al. *An Update on Retinal Prostheses*. 131 Clin Neurophysiol. Jun., 2020; (2019).

[39] *See id.:* Mohsen Farvardin, et al., *The Argus-II Retinal Prosthesis Implantation; From the Global to Local Successful Experience* , 12 Front. Neurosci., 04 September 2018 Sec. Neural Technology (2018).

[40] Maurice Ptito, et al. *Brain-Machine Interfaces to Assist the Blind*, 15 Front. Hum. Neurosci. 9 Feb. 2021

Once any such device is surgically implanted in a person and becomes an integral part of their physical functioning, it would intuitively count as a violation of that person's bodily integrity and liberty for a government actor—or anyone else—to intentionally undercut the functioning of such a device even if they can do so without physical contact to the body of the person who relies on it. As Woodrow Barfield and I have noted in discussing the constitutional law implications of BCI technology, disrupting the operation of such technology with malicious software will just as clearly be a violation of a person's right to personal integrity as a physical attack would be. So would action that disrupts the operation of implanted BCI with electrical impulses. Courts don't need to recognize a specific right to see or sense in order to reach that conclusion: The right to personal integrity and autonomy discussed earlier already embraces it.

Why then might we need a right to see or sense with technology that goes beyond the constitutional (and other legal) protection we already have for bodies? There are a number of reasons we may have a need for such an expanded right to see—and why we may have to think about one that is rooted in freedom of thought or "cognitive liberty" and not just in bodily integrity and freedom. More specifically, there are at least two questions we may have to ask in analyzing whether emerging technologies of seeing are shielded by a constitutional right against government interference and, if so, how this shielding works.

First, some of the technology we see with isn't *integrated* into our body in the same sense as a pacemaker, a prosthetic limb, a brain implant, or an artificial eye surgically embedded in our head. The technology remains outside of our body and is separable from it—even if we rely upon it heavily in the way we regularly perceive our environment. Think again about smartphones and the Supreme Court's statement in *Riley* that the "pervasive and insistent part" they have come to play in "daily life" might make it seem to an observer that they are an "important feature of human anatomy."[41] Smartphones are, of course, not built into our bodies. We can lay them aside and our bodies will keep functioning. In fact, there are movements now to prohibit smartphones in secondary schools and, among some adults, to create smartphone-free periods in their lives, to free themselves from the temptations (and distractions) we

---

[41] Riley v. California, 573 U.S. 373, 403 (2014).

now can't easily avoid on these devices.[42] Yet that doesn't mean that individuals who move in professional or social environments that rely upon these phones could easily dispense with using them. Nor could those who use reminder programs or note-taking software on phones to compensate for natural limits in their memory.

How then should we analyze technologies of seeing that are not integrated into our bodies but perform functions that are arguably crucial to the perceptions we have? They might come in the form of wearable devices, such as virtual reality or augmented reality visors. Or they might consist of other technologies, such as smartphone or computer programs and screens that allow us to watch recordings or live streams of far-away events.

Second, when does a right to see cover technologies that don't merely *replace* perceptual abilities we once had but give us new perceptual powers? If the right to see begins with a right to use our natural biological mechanisms to see—free of state interference—it makes sense that it would cover technologies that are integrated into, or replace, these mechanisms. But we might ask why such a right to bodily integrity and autonomy would cover perceptual abilities that our body never had. There are a number of possible answers to this worth exploring. One is that how we classify a capacity as "natural" or "artificial and extraneous" might change as social and technological conditions change. What counts as necessary may change as we supplement the powers our body has given us with tools that give us (and many others) new powers. Smartphone programs might not only replace natural memory capacities we have lost by allowing someone who has developed difficulty remembering names or other facts to quickly retrieve them on a phone. They might also give people memory capacities our brains never had. And it is worth analyzing whether and when these *new* capacities become part of the right. Is that something that occurs when most people routinely exercise these

---

[42] *See* Phone-Free Schools Movement, at https://phonefreeschoolsmovement.org/; Rebecca Onion, *Will the Phone-Free Movement Work?* Mar. 29, 2024, at https://slate.com/human-interest/2024/03/jonathan-haidt-smartphones-social-media-teens-mental-health.html; Susan Linn, *The Movement to Free Schools of Smartphones Is Winning*, Dec. 13, 2024, at https://prospect.org/education/2024-12-13-movement-free-schools-of-smartphones-winning/; Jonathan Haidt, *The Anxious Generation: How the Great Rewiring of Childhood Is Causing an Epidemic of Mental Illness* (Penguin Press 2024); Stacy Torres, *Opinion: I've Lived into My 40s Without Ever Owning a Smartphone. Hopefully I'll Never Have To*, LA Times, May 27, 2023.

capacities? Is it only true if they *rely* on these capacities to serve certain crucial functions? Or is it social convention—and not the inherent nature of our relationship to the technology—that determines (at least in part) where a technological augmentation of a capacity becomes part of what is protected by a right? As indicated by the smartphone example just considered, these are questions that can be asked about human capacities, as a general matter, but my focus here will be on powers of perception. When do we have a right to see in ways we could never see before?

In asking the latter question, we might ask if the harms that new technologies of perception cause or threaten—most notably the kind of privacy or deception harms discussed more fully in Chapter 5—either disqualify use of such technologies from being covered by a constitutional right or weaken the protection that it offers. Even if a certain kind of perceptual enhancement plays an important role in our lives, if it threatens much more harm to others (or possibly to ourselves) than our natural vision carries, this may justify different constitutional treatment of such technologies. Understanding the implications of these harms or risks of harm may require also thinking more carefully about the underlying interests that new technologies of seeing serve. Is it the same interest that is protected when the law protects our natural vision—or, more generally, protects us against interference with the capacities our bodies give to us? Or are there other underlying interests safeguarded by a right to perceive with the aid of technology?

Extended Perception and Extended Mind

How then might a right to the inviolability of our person—and the visual system that is a part of it—lay the groundwork for protecting *not only* technologies for seeing and mental processing that are integrated into our bodies (in a manner analogous the way a pacemaker become a part of our heart's functioning)—but also for technologies we use outside of it? When will it protect our use of glasses, binoculars, and telescopes? Or computer-connected visors that let us see with "extended reality" technology that replaces or supplements the world in front of us with other objects that are actually far away or nowhere at all?

We might find a starting point for answering such questions in the ethical implications some have drawn from the concept of the "extended mind" set forth by Andy Clark and David Chalmers.[43] The mind, they claim, should be understood not merely as embodied in brain processes but rather as something that includes physical activity that extends outside of our brain and body and into our environment. We often think, in other words, not only with the brain activity that gives rise to and occurs with our mental processes but also by using slide rules or computers to solve a mathematical problem, drawing a sketch, or building a model to help design an environment, or writing a verbal description of our ideas in a notepad. That conduct in the external world isn't merely a support for our thinking. It sometimes *is* thinking—thinking that occurs with tools external to our body and not solely with brain activity inside it.

A core part of their argument is an analogy they draw between the role such uses of tools play in mental processes and the role our brain activity plays: They imagine a person, Otto, whose memory has been weakened by early-stage Alzheimer's and constantly uses a journal to remember addresses and other fact he can no longer recall without help from a written reminder. If he wishes to meet a friend, Inge, at the Museum of Modern Art in New York, he will—upon learning the address of the museum—write it down in his journal, since that is his only hope of remembering it. Inge's memory process may well work differently: Unafflicted by Alzheimer's, she can recall the address without use of a notepad. She can simply store and retrieve it from her natural memory.[44] It doesn't make sense, Chalmers and Clark say, to say that Inge's retrieval of the museum address from her natural memory is wholly an operation of the mind but that Otto's retrieval of it from his journal is not: The journal is as integral to Otto's regular means of remembering facts as Inge's exercise of her unaided short-term memory is to hers. This is an example of what they term "the parity principle": We should count as cognitive in the outside world whatever processes one would count as cognitive in one's head.[45]

Their claim, and the analysis they offer to support it, analyzes the concept of "mind" and whether and when that concept embraces activity conducted outside of our head. However, they do note a possible ethical

---

[43] Andy Clark and David Chalmers, *The Extended Mind*, in Andy Clark, *Supersizing the Mind: Embodiment, Action, and Cognitive Experience*, app. at 220–32 (2008).

[44] *Id.*

[45] *Id.*

implication of this expansion of what counts as part of the "mind" to include certain uses of external thinking tools. "It may be... that in some cases, interfering with someone's environment will have the same moral significance as interfering with their person."[46] If a part of someone's environment is something they've made an essential part of their thinking process, then depriving or interfering with it damages that thinking process as surely as would compelled surgery that alters, or other harm an external actor imposes on, their brain's natural functioning. Other writers have elaborated upon that ethical argument: Neil Levy, for example, defends an "ethical parity principle," whereby "[u]nless we can identify ethically relevant differences between internal and external interventions and alterations [to the way our mind works], we ought to treat them on a par."[47] One implication of this principle, as he notes, concerns privacy practices, and as Clark and Chalmers do in their Otto and Inge example, he uses the example of a journal or notebook: "[I[f it would be wrong to read [a person's] mind because it would be an invasion of their privacy, then it might be equally wrong for the same reason to read their diary." Other writers have considered how damage to computers can be understood as damage to a person: J. Adam Carter and S. Orestis Palermos, for example, ask if and when damage to a person's computer should count as an "assault" on their person.[48]

This kind of ethical framework built around the extended mind argument provides a possible foundation for protecting not only extended contemplation or cognition but also extended perception. If the role a certain technology of perception plays, even when it is separate from our body, is analogous to the role played by the natural mechanisms of perception integrated into our body, then we might have just as much right against interference in it. If interference with our eyesight or brain's visual processing system would violate our rights, so too would action that was functionally equivalent—even though its target was our environment rather than our physical person. This would be true, for example, when someone steals a person's glasses knowing that their vision is too poor

---

[46] *Id.*

[47] Neil Levy, *Neuroethics: Challenges for the 21st Century* 60, 62 (Cambridge University Press 2007).

[48] J. Adam Carter and S. Orestis Palermos, *Is Having Your Computer Compromised a Personal Assault?: The Ethics of Extended Cognition*, 2(4) Journal of American Philosophical Association: 542–560 (2016).

without them to drive, identify people only a few yards in front of them, or perform other basic tasks. Or imagine that a person whose prosopagnosia prevents them from identifying people by their facial appearance wears an augmented reality visor that uses a computerized facial recognition program to verbally inform them of the identity of a friend or family member. Destruction of that device by malicious officials or private actors would have an effect similar to disrupting the natural brain operations of a person without prosopagnosia. As Steven Mann notes in describing the way wearable computers could be used to supplement and enhance perception, this technology can be a valuable tool to "assist partially sighted individuals:" Such technology "computationally augments, diminishes, or alters visual perception in day-to-day situations" to enable them to perceive what they are naturally unable to see.[49] Why then shouldn't a right that protects individuals' capacity to perceive their environment protect such perceptual technology just as it protects the natural visual system?

I have described such rights-based protection for extended perception as being *analogous* to protecting our "extended mind." But we can also understand the protection of extended perception as part and parcel of protecting our extended mind. In fact, this is one justification Seth Kreimer gives for finding that the First Amendment should protect recording. Citing Chalmers and Clark, he says, "[r]ecorded images can serve the same function" as journals we use to preserve memory and that image capture might thus sometimes be counted as part of "an extended cognitive system."[50] And as noted earlier, perception is as much a matter of the brain's processing of sensory information as it is a sense organ's receipt and intake of stimuli. The biology that underlies our sensory perception is not only about receiving inputs for our mental activity but also part and parcel of the operation of that mental activity. So technology that extends our perception, in some cases, is an extension of mental processing as well.

---

[49] Steve Mann, *Wearable Computing: A First Step Toward Personal Imaging*, Computer 30(2), Feb. 1997: 25–37, 28–29.

[50] Seth F. Kreimer, *Pervasive Image Capture and the First Amendment: Memory, Discourse, and the Right to Record*, 159 U. Pa. L. Rev. 335, 380 (2011). *See also David J. Chalmers, Reality+: Virtual Worlds and the Problems of Philosophy*, 295, 299 (W.W. Norton 2022) (arguing that AR technology can "take over the functions of the mind" and thus function as a "mind extender.")

This then provides one basis for understanding constitutional protection for a right to see or sense as a right to see or sense not only with (1) the biological processes nature has given us to perceive the world and (2) technology that has been integrated into our sensory organs or brains to repair the sensory process but also (3) technology that we use *outside of our bodies* in a way that makes it an integral part of our perception.

What constitutional text would provide the basis for treating this kind of a right to see with technology as a constitutional right? Earlier in this chapter, I noted that case law has interpreted a number of rights within the constitution as shielding the integrity of our person (and personal space). The liberty protection of the due process clauses in the Fifth and Fourteenth Amendments does so. They have been understood by courts to shield a "zone of privacy" or "right to be let alone" that the state intrudes into when it constrains our physical liberty or intrudes into our bodies. That protection, I suggested, may protect our freedom of thought as well as our freedom of movement—and freedom against bodily intrusion—since interference with or manipulation of our brains (or nervous system) more generally is an intrusion into our bodies. The same is true of interference in any other organ that is central to our perception of the world: Our eyes, ears, or skin, for example. Analyzing what is encompassed by a concept of "personal sovereignty" or autonomy, Joel Feinberg notes that it includes bodily sovereignty.[51] It also includes, he goes on to argue, much more than that. But at a minimum, it protects our bodies against interference and reserves them for our control. This then is one basis for extending the liberty protection of the due process clauses to technology which, even when it is outside of our bodies, is an integral part of a function that is unquestionably protected when it is performed entirely within our brains and sensory organs. If the technology that played the role of a permanently-implanted cardiac pacemaker could somehow be worn outside of a person's skin instead of being lodged in in their heart, an intentional disruption of it would still interfere (impermissibly) with the heart's rhythm. Similarly, when the technology we rely upon to think and see with is outside of our bodies—and, unlike a pacemaker, it often *is*, when it takes the form of a journal, a set of lenses, or perhaps a wearable computing device—then the government's disabling or restriction of that technology should face intense judicial skepticism

---

[51] Joel Feinberg, 3 *The Moral Limits of the Criminal Law: Harm to Self* 53 (1986).

under the Fifth and Fourteenth Amendments. Government should likewise face that skepticism when, in the exercise of its power to investigate, thwart, and punish criminal activity, it insists upon searching or seizing such technology. Such a search or seizure would not merely be of an "effect" that a person has—it would be a search or seizure of a "person" that threatens the most powerful privacy interest a person has.

Because interference with technologies essential to perception necessarily interferes with a core mental operation, we might also characterize it as a violation of our right to freedom *of thought*—which, the Supreme Court has said, is a core First Amendment right. The Supreme Court said in *Stanley v. Georgia* that it is clear that our "constitutional heritage" "rebels at the thought of giving government the power to control men's minds."[52] Other courts have noted that the government is not permitted under this principle of "freedom of mind" to "scree[n] certain types of stimuli from flowing to a citizen" to restrict their thought.[53] Thus, where government restricts a certain technology of seeing and this interferes with our perception as much as would an unconsented-to medical intrusion into our visual processing biology, it also interferes—one might argue—with our First Amendment "freedom of mind." Indeed, the Supreme Court's landmark freedom of thought ruling in *Stanley v. Georgia* concerned punishment for his *viewing* of something (namely, a film with obscene pornographic content).

There is, however, an obvious question that this argument based on the "extended mind" concept raises: We take all kinds of actions to support our thinking, for example, to give it perceptual inputs it wouldn't otherwise have—actions that affect what we can see or sense. It is not only with a wearable augmented reality device that a person may be able to see in a way they are powerless to see without the technology. We will also see what we otherwise cannot when we pilot a drone carrying a camera over someone else's property—or when we build, or obtain permission to climb, a tall building or tower equipped with a telescope or other device for magnifying distant objects. An even larger range of technologies can shape our thinking: We might calm ourselves by silently reminding ourselves that a problem is not as daunting as it seems but we might also do so with a sedative, by listening to calming music, or

---

[52] Stanley v. Georgia, 394 U.S. 557, 566 (1969).
[53] Doe v. City of Lafayette, Ind., 377 F.3d 757, 765 (7th Cir. 2004).

by doing physical exercise. It seems unlikely, however, that *all* of this conduct counts as an exercise of our mental power that is therefore shielded from government interference by the Constitution's protection for freedom of thought. As the Seventh Circuit Court of Appeals has observed in rejecting a freedom-of-thought claim from a sex offender banned from visiting a city park, "[t]hought and action are intimately entwined; consequently, all regulation of conduct has some impact, albeit indirect, on thought."[54]

In the article in which they proposed the concept of the "extended mind," Chalmers and Clark answer a similar question about limits and ask how far the mind should be extended. One general limit is their claim that thinking of the mind as extending into the environment is justified by the fact that certain "relevant external features" of the environment are "coupled with the human organism." Not every part of our environment can count as "coupled" with the organism. They propose four criteria that they argue justify giving Otto's journal this status and may justify treating other tools of thought that way. A journal a person always relies upon to remember facts is (1) a "constant" feature in the life of the person who is using it, (2) "directly available without difficulty," (3) involves "automatic endorsement" of the notes taken in it, and (4) has been "consciously endorsed" at some point in the past.[55] In their discussion of whether attacks on computers can be attacks on persons, Carter and Palermos similarly write, drawing upon "the mathematical framework of dynamical systems theory" that, in order for a system outside the body to be part of an extended process without us, it is necessary that "the contributing members (i.e., the relevant cognitive agents and their artifacts) interact continuously and reciprocally (on the basis of feedback loops) with each other."[56] On these accounts, there are limits on what kinds of tools can count as a part of our thought—and (on Carter and Palermos's account) be shielded by legal protection for our persons.

There might be similar limits on what kind of technologies of perception are shielded by a *right* of perception. An augmented reality visor or phone application used by someone with prosopagnosia—which, as noted earlier, is a variant of agnosia in which one cannot identify faces—might

---

[54] *Id.*
[55] Clark and Chalmers, *The Extended Mind*, *supra* note 43, at 231.
[56] Carter and Palermos, *The Ethics of Extended Cognition*, *supra* note 48, at 10.

be the kind of perceptual technology that such a person constantly and easily uses to visually identify friends and relatives, trusts in to correctly identify them, and has decided they want to and should use.[57] In these cases, one would likely find the kind of reciprocal interaction—on the basis of a feedback loop—between the biological person and the technology they are using to engage in perception. By contrast, a tower that a person climbs up to get a better view of a certain landscape or a drone they operate occasionally when they need an aerial vantage point might not count as an integral part of perception under these accounts. As noted in the next two sections, however, even when technologies are not integral to seeing, there may be reasons to include them - at least in some circumstances—within the kind of right to see that this book discusses.

The Right to Enhanced Perception in Unfamiliar Jurisprudential Territory

The key hypotheticals discussed above presented examples of situations where a person uses technology to replace a perceptual or other cognitive power that has deteriorated or is normally present (without technology) in others: Otto's journal serves as a *replacement* for certain aspects of his Alzheimer-weakened memory. An AR application that allows a face-blind person to recognize faces is a *replacement* for a perceptual task that is normally performed with a person's brain processing and natural vision but, in this person's case, requires technology outside the body. The argument explored above is that, if such a replacement for a process that counts as thinking or seeing when performed with a person's natural sensory and cognitive processes also counts as such when performed partially with external tools, then those external tools are part and parcel of the thinking or seeing process. As such, they are covered by a right (constitutional or otherwise) that protects that activity. Even though they are, unlike a pacemaker or implanted brain chip, outside the body, they are essentially a part of the person—and covered by the rights-based shield that protects the person from external interference.

That analysis, however, leaves another question unanswered: What should the constitutional rules be when our use of technology for seeing (or thinking) doesn't simply replace a part of our natural thinking but

---

[57] *Id.*

augments it? Recording an event, for example, doesn't merely allow us to reexperience it and think more carefully about it in precisely the same way we would if we could only do so by conjuring up a memory of it. It allows us to repeatedly watch a record of the event and see details we likely couldn't have recalled without the recording. It is true that we can often pull more details from a past memory into our consciousness with repeated efforts to recall them. It is also true that video recordings are limited by the angle they were shot from and by certain conditions (such as dim lighting). But recording still generally allows individuals to remember more detail than they otherwise could—and to learn new details about even events they have personally witnessed.

Other enhancements of vision are less familiar. Thermal imagers allow those who have them to see images built from infrared radiation that our eyes normally don't register. As noted in Chapter 1, artificial eyes and visual processing brain-computer interfaces might allow someone to see this kind of normally invisible light all the time—and perhaps with artificial organs built into their bodies rather than a device (like a thermal imager) that they carry with them.

When, if ever, are such technological *enhancements* to our seeing protected by a right to see with technology? If and when they are protected constitutionally—on what grounds are they protected? Would embedding any such device into our body automatically make it a protected part of our person no matter what it can do? Would it obtain such protection where a person shows they constantly rely upon it, can always access it without difficulty, trust in it produce accurate information, and endorse it?

There are two reasons some might at least tentatively respond by answering "no." One is that the set of perceptual powers that we naturally have might serve as a benchmark of sorts for understanding what minimal conditions are needed for individuals to exercise the autonomy they are assured of in the American constitutional system. Individuals have a right against physical interference with, or restrictions that deny them, free use of their arms (or at least uses that don't harm others). That may mean that they have the same right with respect to prosthetic arms. It doesn't necessarily mean that anyone would have the right to produce or use a mechanical exoskeleton that endows them with superhuman strength. The latter, one might argue, is not necessary to an autonomous existence in the way that our familiar type of freedom to engage in physical movement is. For similar reasons, while being able to see (with natural eyes or artificial eyes that replace them) may be shielded against state interference, that does not mean that X-ray vision should be covered by the same shield. In the social world and built environment of the current era, an existence characterized by individual autonomy often requires that we are able to perceive our surroundings but not that we be able to see through walls.[58]

Of course, conditions can change. What is sufficient for the exercise of autonomy at one point in time may not be sufficient a couple of decades later. The uses people make of smartphones may have seemed like an unnecessary luxury had they been available in the 1990s: It would have seemed odd to say anyone had a right to carry a camera, a GPS device, and 128 Gigabytes of computer storage with them. Now, a law severely restricting use of smartphones would, for many individuals, prevent them from performing activities essential to their work or family lives.

A second reason that people may hesitate to extend a right to see to cover all forms of enhanced seeing lies in the new risks of harm some enhancements necessarily or likely generate. X-ray vision or other kinds of "see-through perception" that allows individuals to peer through walls would, of course, raise significant threats to the privacy of the home. That is why the Supreme Court found that it was unconstitutional for police to generate an image of a home's interior from the street outside using a thermal imaging device—unless they first obtained a warrant from a neutral magistrate.[59]

Such harms might merit different treatment for a technological *enhancement* of vision even if the technology is integrated into one's body. A prosthetic arm that is designed to fire laser beams (capable of blinding or burning) would likely be subject to greater regulation than an ordinary prosthetic arm. So would an artificial eye that allows the individual possessing it to see through walls. Even if a person came to make regular use of such capacities, the privacy harm they raise for others may require and justify regulation of a kind that would be impossible if such an artificial eye were given the same shielding as our person normally receives against interference. As I have written before, this may be true even of technology that allows us to generate inner thoughts in a certain way: If a defamatory falsehood or statement inciting a crime is delivered to

---

[58] Lawrence Lessig, *Constitution and Code*, 27 Cumb. L. Rev. 1–15 (1997).
[59] *Kyllo*, 533 U.S. at 36.

others not through verbal expression but in thoughts that brain-computer interface technology one day makes shareable *without* spoken or written expression, such shared thought may still cause the same harms as the kind of defamation and incitement that has for decades been subject to liability. Consequently, even when a technology that augments our perception or thinking power *is* linked to our thinking in such a way that it may be justifiable to say that our use of it *constitutes* "seeing" or "thinking" (under criteria similar to those discussed earlier in work such as that of Chalmers and Clark or Carter and Palermos), that doesn't necessarily mean that has to be insulated against legislative or regulatory limits to the same degree as our natural thinking and seeing. The harms accompanied by the technology arguably justify different constitutional rules.[60]

Recording technology can raise this kind of concern. A federal appeals court recently found that it was permissible under the First Amendment for a state (Texas) to ban use of a drone to capture pictures of private property except in limited circumstances.[61] The use of Google Glass and other glasses or visors that come equipped with the capacity to record everything that occurs in front of a person has likewise raised privacy concerns. In fact, while augmented reality with facial-recognition capacities may be integral to seeing for individuals with prosopagnosia, such technology could also potentially be used to eliminate anonymity in public settings.[62]

There are, however, at least two reasons we should not therefore conclude that enhancements to our senses are *automatically* disqualified from rights protection. One is simply that what is true for some kinds of enhancements might not be true for others. Even if *some* kinds of enhanced vision are unnecessary for individuals to retain their freedom

---

[60] This is a question I have explored in Marc Jonathan Blitz, *Freedom of Thought and the Structure of American Constitutional Rights* in ed. Marc Jonathan Blitz and Jan Christoph Bublitz, *The Law and Ethics of Freedom of Thought, Volume 1: Neuroscience, Autonomy, and Individual Rights* (New York: Palgrave Macmillan; 2021).

[61] National Press Photographers Ass'n v. McCraw, 90 F.4th 770 (5th Cir. 2024).

[62] Given such potential harms, it is perhaps not surprising courts have found the First Amendment doesn't bar state laws that restrict creation of "face prints" for facial recognition systems. *See In re Clearview AI, Inc., Consumer Priv. Litig.*, 585 F. Supp. 3d 1111, 1119–1121 (N.D. Ill. 2022). For a discussion of First Amendment questions surrounding use of a device that "records everything a person does" see Jane R. Bambauer, *Glass Half Empty*, 64 UCLA L. Rev. Discourse 434, 439 (2016).

of thought or physical freedom, others might be. Even if some new technologies for seeing create harm, or risks of harm, that the government should have leeway to guard against, *other* types of enhancement may be free of them. Indeed, it is even possible that the version of sense-enhancing technology designed in 2040 may have safeguards included that mitigate the harms that existed in the 2030 design. If that is true, the same technology may merit different legal treatment in different years. Or courts may feel that it is necessary to generalize and build a legal rule that will work for plausible worst-case scenarios even if they are not facing one. In its analysis of thermal imaging devices, for example, the Court in *Kyllo v. United States* took the position that although the infrared technology used in the case itself was "relatively crude... the rule we adopt must take account of more sophisticated systems that are already in use or in development."[63]

Consider an example. The capacity of extended reality technology to give me telepresence—to feel as though I am at a different location—may be an unfamiliar one. It may be one that Americans and individuals in other countries have long been able to do without. It may cause harm when I am given telepresence in a house I don't have permission to enter. But that doesn't mean that telepresence is not deeply valuable in my exercise of perception, and harmful or problematic, when it virtually transports me into a historical structure thousands of miles away that I *do* have a right or permission to enter. Why should the law be able to ban a form of high-tech perception on the ground that some variants of it are without value or come with danger to others?

There is also a second reason that we should perhaps have a right to see with technology even when it lets us see in ways human beings have not been able to see before. Even when a certain kind of conduct threatens harm that does not mean that it can't covered by a right. Consider the speech protected by the First Amendment: There are certainly many harms that speech can cause. Two of them (defamatory damage to reputation and a crime resulting from incitement) were just mentioned. In those two cases, the harms raised have led America's legal tradition to withhold First Amendment protection from those two categories of speech. In other cases, speech remains protected in spite of the harms it may cause: The expression shielded by the First Amendment can lead listeners

---

[63] *Kyllo*, 533 U.S. at, 36.

to support (and perhaps successfully push for the enactment of) harmful policy. It can cause deep offense or hurt feelings. Commercial speech can cause consumers to purchase products they later regret purchasing.

Similarly, it may be that some of the ways individuals learn about their natural and social environments by observing come with at least some cost to privacy—but that the Constitution should protect them anyway to assure that these avenues generally remain open for individuals to enlighten themselves, about public affairs, about personal interests in their natural or urban surroundings, or about facts that have deep personal significance (such as a place that is connected to their personal history). Moreover, in some cases, such observations are only possible with technologically enhanced forms of seeing. Sometimes our only alternative to learning about information in the secondhand reports of others—our only way of seeing something important to us with our own eyes—requires seeing it with the aid of advanced cameras or computer technology. There are parts of the world we may be able to see only in the camera feed we receive from a drone or with a camera placed there by a local resident. There are environments we may be able to immerse ourselves in only with virtual reality technology—and the telepresence it might give us in a far-away (or now vanished) place. There are cities we may be able to explore only by using Google Streetview or some other global mapping system that allows us to see and wander through places to which it is impractical for us to travel. (Thanks to the company, Mapillary, and some photo takers in Cuba, Louis Zemel might even have had the chance, today, to visually explore Havana, Cuba, in a way that was impossible in the 1960s.)[64] Rather than evicting this kind of perception from the territory of a right to observe the world, courts might instead do what they do for other rights: Recognize a *qualified* right that allows room (at least in some circumstances) for the government to protect privacy and security interests that need protecting. In the realm of free speech rights, for example, courts protect speech that the government wants to restrict because of objections to its political or other content, but it subjects to only "intermediate scrutiny" laws that place limits on speakers to protect the flow of traffic in streets or the ability of all individuals to use a public park. Similarly, a right to see with technology may well—as I have argued

---

[64] *See* Mimi Whitefield, *3-D images will allow users to get virtual view of Havana streets*, Miami Herald, Sep. 26, 2016, at https://www.miamiherald.com/news/nation-world/world/americas/cuba/article104007501.html#storylink=cpy

before[65]—be subject to such intermediate scrutiny that allows room for laws protecting individuals' privacy or safety.

In Fourth Amendment jurisprudence, our right to privacy from intensive government surveillance has force in homes—which, since they are at the "very core of Fourth Amendment protection," is presumptively off-limits to law enforcement officers unless and until they receive a warrant based on "probable cause." Yet our right to privacy continues to have force even outside this Fourth Amendment "core" inside of our homes—in the public space outside of it. In public highways, airports, schools, and workplaces, the public unquestionably has a powerful interest in receiving protection against threats to safety and law enforcement has a duty to provide that protection: It has a duty to try to keep drunk drivers off of roads, detect and thwart terrorism threats directed at air travel, and keep schools and workplaces free of dangerous drug use. But that doesn't mean that Fourth Amendment privacy protections simply disappear in those places. It means instead that they must work differently—in a way that lets courts balance privacy and security interests and leave room for officials to sometimes engage in random or pervasive suspicionless searches (in airport searches for weapons, for example, or random workplace testing for drugs). This too can provide a model of sorts for courts to strike a balance between individuals' right to observe the world and others' right to remain free of surveillance by their fellow citizens.

A Summary of a Constitutional Framework for Perception Based on Personal Autonomy

This chapter has made three points that should help us begin to define the contours of a right to see with technology on constitutional terrain less familiar than the First Amendment ground explored in the last chapter. First, the constitutional right that is most suitable to protecting our use of natural vision is the autonomy right built into protection for bodily and mental integrity. My argument is not that the First Amendment protection for the right to record is without value. Far from it. Those cases have done essential work in explaining how First Amendment doctrine developed primarily for verbal communication—and to a lesser extent, for

---

[65] See Blitz, The Right to Map, supra note 3, at 190–201.

symbolic and artistic expression—should apply to a key part of communication that occurs in the twenty-first century: The exchange of images and video and audio recording. In doing so, they protect a certain kind of enhanced seeing that we can exercise in collaboration with others, jointly engaging in a process of First Amendment communication. However, there is a value in perception that this account fails to capture and we can begin to take stock of it by focusing instead on the integrity and autonomy protections that one finds in the Fourth, Fifth, and Fourteenth Amendments. These protections generally shield not only the biological machinery we were born with for seeing the world, but artificial machinery implanted in us to restore or take over some of the functions of our natural organs.

Second, using the extended mind theory—and particularly, the ethical arguments some have constructed around it—as a template, we might understand protections of our personal integrity and autonomy to cover not only the seeing we do with natural organs or direct replacements but also the seeing we do with tools that play a role necessary for us to successfully perceive the world. That these tools are outside of our bodies and don't have a form identical or similar to a natural sensory organ doesn't exclude them from being covered by the right: Just as a journal that has replaced natural memory is starkly different in form and the way it works from the recall processes that occur when neurons fire, so an AR visor that identifies faces by displaying words on the inside of the visor or that uses a similar mechanism that compensates for attention problems is quite different from ways that a person's brain would allow for recognition of faces or more successful forms of focusing on the environment.

The third key point made by the chapter is not technically a claim. It is rather a tentative set of suggestions about what courts will have to think about when they turn from assessing technologies that replace lost functions to those technologies that endow us with new capacities. As I've said, courts should examine whether such technologies are as central to our autonomy (or some other interest underlying our freedom of body and mind) as technologies that restore old functions instead of giving us new ones, whether they generate harms (to the perceiver or others) that are substantially more concerning than the harm likely to arise from seeing with our natural visual systems, and how social and technological developments might affect or alter the answers for these questions.

References

Articles and Books

Adam Z. Consciousness: A User's Guide, 208–223, 377 (Yale University Press 2002).

Lauren N. Ayton, et al., *An Update on Retinal Prostheses*. 131 Clin Neurophysiol. Jun, 2020. (2019).

Jane R. Bambauer, *Glass Half Empty*, 64 UCLA L. Rev. Discourse 434 (2016).

Marc Jonathan Blitz, *Freedom of Thought for the Extended Mind: Cognitive Enhancement and the Constitution*, 2010 Wis. L. Rev. 1049 (2010).

Tamara Bonaci, Ryan Calo, and Howard Jay Chizeck, *App Stores for the Brain: Privacy & Security in Brain-Computer Interfaces*, Proceedings of the IEEE 2014 International Symposium on Ethics in Engineering, Science, and Technology (2014).

J. Adam Carter, *Varieties of (Extended) Thought Manipulation*, in ed. Marc Jonathan Blitz and Jan Christoph Bublitz, The Law and Ethics of Freedom of Thought, Volume 1: Neuroscience, Autonomy, and Individual Rights (Palgrave Macmillan 2021).

J. Adam Carter and S. Orestis Palermos, *Is Having Your Computer Compromised a Personal Assault? The Ethics of Extended Cognition*, 2(4) Journal of American Philosophical Association: 542–560 (2016).

Andy Clark and David Chalmers, *The Extended Mind*, in Andy Clark, *Supersizing the Mind: Embodiment, Action, and Cognitive Experience*, app. at 220–32 (2008).

Allen Coin and Veljko Dubljević, *An Introduction to Policy, Identity, and Neurotechnology: The Neuroethics of Brain–Computer Interfaces*, in ed Allen Coin and Veljko Dubljević, *Policy, Identity, and Neurotechnology: The Neuroethics of Brain–Computer Interfaces* (Springer 2023).

Subrat Das, et al., *Cybersecurity: The Need for Data and Patient Safety with Cardiac Implantable Electronic Devices*, 18(3) Heart Rhythm (March 2021).

John E. Dowling and Joseph L Dowling, Jr. *Vision: How it Works and What Can Go Wrong* (MIT 2016).

Mohsen Farvardin, et al. 12 Front. Neurosci., 04 September 2018 Sec. Neural Technology (2018).

Joel Feinberg, *The Moral Limits of the Criminal Law: Harm to Self* (Oxford University Press 1986).

Jonathan Haidt, *The Anxious Generation: How the Great Rewiring of Childhood Is Causing an Epidemic of Mental Illness* (Penguin Press 2024).

R. A. Juskalian, New Implant for Blind People Jacks Directly into the Brain, MIT Technology Review, Feb. 6, 2020, at https://www.technologyreview.com/2020/02/06/844908/a-new-implant-for-blind-people-jacks-directly-into-the-brain/.

Margot E. Kaminski, *Drone Federalism: Civilian Drones and the Things They Carry*, 4 Cal. L. Rev. Circuit 57 (2013).

Seth F. Kreimer, *Pervasive Image Capture and the First Amendment: Memory, Discourse, and the Right to Record*, 159 U. Pa. L. Rev. 335 (2011).

Lawrence Lessig, *Constitution and Code*, 27 Cumb. L. Rev. 1 (1997).

Margaret S. Livingstone, *Vision and Art* (Updated and Expanded Edition) 24 (Abrams 2014).

Neil Levy, *Neuroethics: Challenges for the 21st Century* 60, 62 (Cambridge University Press 2007).

Susan Linn, The Movement to Free Schools of Smartphones Is Winning, Dec. 13, 2024, at https://prospect.org/education/2024-12-13-movement-free-schools-of-smartphones-winning/.

Jamie Luguri and Lior Jacob Strahilavitz, *Shining a Light on Dark Patterns*, 13 J. Legal Analysis 43 (2021).

Steve Mann, *Wearable Computing: A First Step Toward Personal Imaging*, Computer 30(2), Feb. 1997.

Rebecca Onion, *Will the Phone-Free Movement Work?* Mar. 29, 2024, at https://slate.com/human-interest/2024/03/jonathan-haidt-smartphones-social-media-teens-mental-health.html.

Maurice Ptito, et al. *Brain-Machine Interfaces to Assist the Blind*, 15 Front. Hum. Neurosci. 9 Feb. 2021.

Stacy Torres, Opinion: I've Lived into My 40s Without Ever Owning a Smartphone. Hopefully I'll Never Have To, LA Times, May 27, 2023.

Mimi Whitefield, *3-D images will allow users to get virtual view of Havana streets*, Miami Herald, Sep. 26, 2016, at https://www.miamiherald.com/news/nation-world/world/americas/cuba/article104007501.html#storylink=cpy

Adam Zeman, *Consciousness: A User's Guide* (Yale University Press 2002).

Other

Blacklist, Season 6, Episode 10 (*"The Cryptobanker"*).
Homeland, Season 2, episode 10 (*"Broken Hearts"*).

Cases

Birchfield v. North Dakota, 579 U.S. 438 (2016).
Cruzan v. Director of Missouri Department of Health, 497 U.S. 261 (1990).
Doe v. City of Lafayette, Ind., 377 F.3d 757 (7th Cir. 2004).

Dobbs v. Jackson Women's Health Org., 597 U.S. 215 (2022).
Irizarry v. Yehia, 38 F.4th 1282 (10th Cir. 2022).
Kyllo v. United States, 533 U.S. 27 (2001).
National Press Photographers Ass'n v. McCraw, 90 F.4th 770 (5th Cir. 2024).
Pottowatomie County v. Earls, 536 U.S. 822 (2002).
Riley v. California, 573 U.S. 373 (2014).
Skinner v. Ry. Labor Executives' Ass'n, 489 U.S. 602 (1989).
Stanley v. Georgia, 394 U.S. 557, 566 (1969).
State v. Villarreal, 476 S.W.3d 45 (Tex. App. 2014)
Terry v. Ohio, 392 U.S. 1 (1968).
Union Pac. R. Co. v. Botsford, 141 U.S. 250 (1891).
Vernonia Sch. Dist. 47J v. Acton, 515 U.S. 646 (1995).
Washington v. Harper, 494 U.S. 210 (1990).
W. Watersheds Project v. Michael, 869 F.3d 1189 (10th Cir. 2017).

CHAPTER 5

Cognitive Liberty, Privacy, and Extended Perception

Abstract This chapter first summarizes (and slightly recasts) the previous chapters by noting that they grounded a right to see or sense the world with technology in two larger constitutional rights: (1) A right to express oneself with technology, for example, by video recording and (2) A right to use one's visual system—including technology integrated into it—behind constitutional shielding for one's personal integrity and autonomy. The justification for the right becomes more tenuous, however, as one moves to forms of perceiving with technology that are more distant from the core of each of these rights. This chapter argues that those who think about perception with technology must therefore emphasize and understand the role of another set of interests that underlie many forms of seeing with technology: Our interest in what many writers call "cognitive liberty."

Keywords Cognitive liberty · Constitutional rights · Interests · Law and technology · Science fiction · Lifelogging · Brain-computer interface device · Virtual reality · Augmented reality · Perception · Freedom of speech · Fourth amendment · Surveillance

Moving Beyond "Core" Territories for a Right to See (in Expressive and Bodily Freedom)

We live in an age where people constantly have computer or phone screens available—and can constantly capture and receive images. We are entering an age where experiencing these recorded perceptions in three dimensions rather than two may become more common. And where individuals—and not just the state that watches over them—can mount cameras on drones and vehicles, record (and store) hours or days of footage, some of it magnified with zoom lenses, and analyze or edit such images with artificial intelligence, or, for that matter, fabricate them from scratch.

The Constitution nowhere expressly gives us a right to observe the world—whether with natural or technologically enhanced vision. It doesn't clearly give us a right to walk around public property and take mental notes by scrutinizing it closely. Nor does it give us a right to enhance our vision, with the help of any of the camera or computer technology I have just described.

So the previous chapters have focused on exploring if there is a right to see (and see with technology) embedded in one or both of two more familiar constitutional rights: (1) a right to speak, to otherwise convey information to audiences, or receive it from others and (2) a right to physical integrity, bodily autonomy, and freedom from physical restraint—presumably insulating against external interference our right to use our visual system as much as any other part of our body. In analyzing both of these rights, however, the case for constitutional protection becomes more tenuous—and raises more questions—as the kind of seeing or sensing that we are doing moves further away from the paradigmatic exercise of the right.

Thus, in Chapters 2 and 3 of this book, I agreed with the many courts that have found that recording video or audio is a kind of speech creation and thus, shielded by protection for speech. But I pointed out that it is harder to justify providing the same First Amendment protection on the same grounds when the way we register images occurs *without* the creation of an image we can share—when we engage in silent observation that leaves its mark on our minds without leaving a shareable record in the world. Does such a right cover our silent sensing even when that sensing is not embedded in a medium of communication? Does it obtain protection simply to the extent it provides us with information, or provides us with the raw material we can later use to generate images in our minds?

The more we turn our focus away from how we communicate information to others and focus on how we perceive and absorb information—the more we focus on *non-expressive* activity that is more likely to cause physical disruption, or intrude onto others' property, than the simple act of sending a video on a text or posting it on a social media site—the harder it is to make a case for robust constitutional protection. We move from a traditional area of robust individual autonomy—the realm of speech—to types of non-expressive conduct which, even when it is invaluable for obtaining information, raise risks of a kind the state has usually been free to regulate (risks to physical safety or others' property rights), without facing significant constitutional obstacles.

The previous chapter raised a parallel question about what happens when seeing moves from a "core" realm where our autonomy interests are at strongest to a periphery where they are weakened, and harm is more likely. In Chapter 4, however, the realm of autonomy in question was not our right to give voice to what we choose to express but rather the right to control our physical self, and to some extent our psychological self—and remain free from state interference as we do so. It is a right to have autonomy in our bodies and minds. Again, however, the case for claiming that seeing or other sensing technology is covered by such a right seems to become more questionable as we move away from the core examples of it. Our use of our eyes seems clearly covered by a right to bodily autonomy. So too is our use of a visual prosthetic that is integrated into our bodies to replace a natural organ (or component of one) that has ceased to function. The same might be true even of a *functional* replacement for natural capacity that we wear rather than surgically implant—such as a kind of artificial eye that is built into a visor that can temporarily connect with our brain's visual system when worn rather than being inseparable from our body. When the technology is not only separable from our body but also gives it capacities of perception it never had before, then the connection seems weaker. Thus, Chapter 4 asked whether the autonomy we exercise over our perception covers our creation of visual reveries that we experience not as an ill-defined image generated in our imagination (or a faded memory) but in the light that a VR system feeds to us. Or the near-perfect copy of a far-away location it creates to give us telepresence in a far-away location. Or certain capacities to see radiation or sound that otherwise cannot be seen. Or a machine that guides our attention in ways we haven't been able to guide it before. Like information gathering that is distant, and perhaps severed, from speech, the tools of perception we'd use to generate such capacities may not only be separate from our bodies and brains—they may also have no functional analogue in our natural physical functioning. Whatever autonomy we had—and enjoyed under the protection of constitutional rights—was exercised in the absence of these new, technologically-generated capacities or any natural analogue of them. So on what basis, one may ask, can we insist they be treated as part and parcel of our person, with legal interference into them treated as unconstitutional interference in our person? Moreover, when we use tools that give us new capacities, these may raise risks of harm that the exercise of our natural powers hasn't generated before, or were much less likely to generate.

Rights, Interests, and Technological Change

In this chapter, therefore, I want to suggest that—in analyzing technologies or seeing—we can benefit from asking additional questions about how to analyze the constitutional status of technologies of seeing that seem, in a sense, to be outside of each of the core rights that I have described above. To do so, I want to briefly digress from the focus on the right to see to ask some more general questions about how courts adapt, and should adapt, a particular right (and the doctrine associated with it) to certain technological changes. More specifically, I want to distinguish between two kinds of legal adaptation to technology, the first much simpler than the second.

First, rights—as we have seen throughout this book—cover certain kinds of conduct. The First Amendment protects a right to *speak*.[1] It also protects a right to form and have *beliefs and other thoughts*.[2] The Fifth and Fourteenth Amendments give us a right to *bodily integrity and*

---

[1] The text of the First Amendment includes language stating that "Congress shall make no law... abridging the freedom of speech." As Chapter 2 explains, the Court has long held that this includes the right to "utter or print" but now also includes numerous forms of artistic expression, such as movies (*see* Burstyn v. Wilson, 343 U.S. 495 (1952) and video games (*see* Brown v. Entm't Merchants Ass'n, 564 U.S. 786, 790 (2011)) and to symbolic conduct (*See* United States v. O'Brien, 391 U.S. 367, 379–80 (1968)).

[2] *See, e.g.*, Paris Adult Theatre I v. Slaton, 413 U.S. 49, 67–68 (1973) (acknowledging and reaffirming the holding of *Stanley v. Georgia*, four years earlier, that the government "has no legitimate interest in 'control [of] the moral content of a person's thoughts" and that "[t]he fantasies of a drug addict" and presumably of any other person "are his own and beyond the reach of government").

autonomy.³ At times, the scope of the right may appear to change fundamentally—but such a change may be superficial. Each of these rights may continue to protect the same category of activity that it has always protected—speech, thinking, bodily movement—but, thanks to technological change, that activity increasingly takes a different form than it has in the past. Consider the right to freedom of speech. Our First Amendment right to speak now protects the way we speak in the twenty-first century, in social media posts, text messages, emails, and video chats, as well as in face-to-face conversations.⁴ Although protecting speech may have once meant primarily protecting only the statements we verbally utter, the words we print on paper, and the art we create on canvases or in other physical mediums, it now necessarily entails protecting digital communications and other digital expression because that is the form our speech now takes.⁵ A First Amendment jurisprudence that failed to adjust to technology in that way would leave much of modern communication with no defense against state censorship. The same might be true, in certain respects, of a right to see. If use of perception-enhancing technologies is as integral a component of how we see the world as social media posts are to how we speak about it, then the Constitution should protect *this kind* of seeing from government interference as strongly as it protects that which we do with our unaided vision. The scope of the right may appear to change. But that is only because the nature of what it protects has been transformed by technology and the right has to cover modern forms of seeing as well as older forms.

The value of such an adjustment may seem obvious: Protection for "freedom of speech" won't deserve that description if it doesn't protect speech in whatever form it takes. However, there is another important

---

³ *See* Terry v. Ohio, 392 U.S. 1, 8–9 (1968) (emphasizing the "the right of every individual to the possession and control of his own person"); *Cruzan v. Director of Missouri Department of Health*, 497 U.S. 261 (1990) (recognizing individuals' right to refuse medical treatment they do not wish to receive).

⁴ *See, e.g.,* Packingham v. North Carolina, 582 U.S. 98, 104 (2017) (noting that "the most important places (in a spatial sense) for "the exchange of views" are now found in "cyberspace—the 'vast democratic forums of the Internet' in general, and social media in particular" and that given the importance of internet communication to modern speech, "the Court must exercise extreme caution before suggesting that the First Amendment provides scant protection for access to vast networks in that medium").

⁵ District of Columbia v. Heller, 554 U.S 570, 582 (2008) (noting that the First Amendment applies to "modern forms of communication" such as the internet).

reason that rights have to adjust to technological change in this way. As Thomas Scanlon notes, many philosophers understand rights as establishing duties that are "justified by the fact that if they are enshrined in law or otherwise recognized they will serve to protect certain *interests*."[6] Our right to freedom of speech, for example, protects our interest in being able to share information of public importance so that we will be well equipped to act as informed voters and citizens. It also protects our interest in learning the truth about the world we live in, without the government steering those intellectual explorations for its own benefit. And it protects our interest in developing ourselves by exploring different ideas—and committing ourselves to those we find valuable.

Technological change, however, can detach the right (as it is defined by legal doctrine) from the interests it is supposed to protect. This will happen, for example, if a right to freedom of speech protects only the old-fashioned version of speech. As speech increasingly takes electronic form, it is that *electronic* speech we use to engage in democratic deliberation by engaging in public discourse on social media and in electronic newsletters. It is in internet searches and electronic books and articles that we increasingly learn about scientific truths or historical facts. It is on the World Wide Web and in numerous electronic conversations—as well as in-person discussions, meetings, and excursions—that we develop personal belief systems and discover which principles we should adhere to and what careers or other activities we will find meaningful. A First Amendment that leaves modern digital communications or explorations without constitutional protection would raise no barrier against state attacks on the way that denizens of the twenty-first century govern themselves and shape their democracies, seek truth, or shape their own belief systems. The right would increasingly become outmoded—protecting only archaic forms of the conduct we take to further these interests.

The same is true of the Fourth Amendment's protection of privacy against government surveillance. We have a powerful interest in the privacy of our communications with family, close friends, and others with whom we have sensitive discussions. The privacy of the home shields that interest—and Fourth Amendment law has long recognized this by treating the home as being at the "core" of the Fourth Amendment's

---

[6] *See* Thomas M. Scanlon, *Rights and Interests*, in eds. Kaushik Basu, and Ravi Kanbur, *Arguments for a Better World: Essays in Honor of Amartya Sen: Volume I: Ethics, Welfare, and Measurement* (Oxford, 2008) (emphasis added).

protection. But the home is not the only site where we engage in private conversations. Such conversations increasingly occur in emails, text messages, and video chats. Nor is it the only place where we keep confidential diaries or other private papers. Many of our most private files are now elsewhere: stored in the memory banks of cell phones or in "the cloud."[7] So Fourth Amendment law can only protect our privacy interests if its protection extends not only to brick-and-mortar environments like the home but also to the digital environments where those interests are now at stake in the twenty-first century.

In other cases, a right has to adjust to technology not by protecting a new high-tech form of the conduct it shields but rather by protecting *against* a new high-tech form of intrusion. For example, in the past, the Court has been sensitive to the fact that government officials can effectively silence us by denying us a space to speak: It would be of small comfort that we have a right to say anything we want if we are barred from saying it public parks or streets where it can reach people. But as we increasingly communicate with people in electronic rather than physical spaces, courts have to be wary of the ways the government might use its power over electronic media to deny us space to speak in that area. Similarly, surveillance technologies change: Thus, the Supreme Court has held that our in-home privacy must not only be safeguarded against the kinds of physical entries into our houses that the Fourth Amendment's framers were most concerned about in the eighteenth century. It must also protect us against "virtual" entries into our homes with high-tech devices, such as thermal imagers that let the government see through walls.[8]

In short, rights protect our interests by protecting certain spheres of our lives against certain threats. To do so, they must protect the form these spheres take in contemporary life (protecting digital speech and electronic private realms, for example) and they must protect against contemporary versions of these threats (by protecting against electronic as well as physical surveillance).

Yet fitting old rights to new technological landscapes isn't always a simple task. At times, technology will generate a gap between a right and

---

[7] That is, a server or set of servers operated by a company that agrees to store our material for us in a way that we can access it over the internet while keeping it confidential from others.

[8] Kyllo v. United States, 533 U.S. 27 (2001).

its underlying interest in a way that doesn't have a clear judicial answer.[9] For example, modern surveillance technology threatens our privacy not only by enabling government to watch us in private physical spaces like the home and private electronic spaces, such as email communications. It also does so in *public spaces*. Increasingly, large stretches of urban life now take place under the gaze of city-wide camera systems. With the rise of pervasive cellphone location tracking and call storage, it is increasingly difficult for individuals to participate in modern life without generating a trail of records about where they have been and who they have communicated with. The possibility that we will be constantly videorecorded by government cameras or followed by location-tracking technology as we move through public streets raises a less familiar kind of threat to our privacy. The technological challenge is not as simple a challenge for courts as the threat presented by transformations of private spaces or of the technological threats to the privacy of these spaces.

This is for a couple of reasons. In the first place, to the extent we have privacy interests in public spaces, Fourth Amendment law cannot protect these privacy interests in the same way it protects them in the home or in a private electronic space—by simply shutting the government out or otherwise ensuring that what we do there cannot be seen or recorded (unless government can show it is a powerful reason to enter into such a private space). Government officials are expected to provide protection for physical safety in public spaces. They are often expected to have a presence there—for example, when there are public events or activities that require the provision of security to protect against physical threats. Moreover, in the public space we share with other individuals, we can't expect that our actions will remain unseen by other individuals. Nor, in a world where cellphone cameras are pervasive, can we insist that we

---

[9] *See* David S. Han, *Constitutional Rights and Technological Change*, 54 U.C. Davis L. Rev. 71, 76 (2020) (arguing that "in moments of significant and rapid technological change like the present, courts should loosen the traditionally strong preference for clear, rule-like approaches in constitutional rights doctrine and embrace an incremental degree of doctrinal complexity and open-endedness"). Another writer who describes a variant of this challenge with respect how technology raises questions about the autonomy of our bodies is Andrea Matwyshyn, who writes that we are "entering a technological age where the line between the human body and the machine is beginning to blur" and notes that "the law is currently unprepared to address the] harms and the social transformation that the Internet of Bodies will occasion." *The Internet of Bodies*, 61 Wm. & Mary L. Rev. 77, 86–87, 90 (2019).

never be recorded in public space—when, for example, someone wants to capture a slice of the urban life that surrounds them and we are a part of the event or scene they photograph. In short, when we bring our interest in privacy into the public realm it must share space with, and be balanced against, *other* interests—such as the public's interest in security and in being able to gather information from the public spaces set aside for unfettered communication and vigorous debate and exploration.

In the second place, privacy and autonomy interests *themselves* may take a different form in our public lives than they do in intimate activity within the home. For example, the privacy we seek in public may be more likely to serve the purpose of shielding, against government monitoring and interference, certain political discussions and other political activity we take with fellow citizens rather than intimate conversations with friends and family members. So courts may have to explore whether the nature and degree of Fourth Amendment protection for that kind of privacy interest is similar to that protection it provides for in-home privacy and other private spaces. If, to some observers, there seems to be a gap between Fourth Amendment coverage and the interest it is supposed to protect, that may be in part because of disagreements or uncertainty about the nature of those interests.

In fact, it is often the case that the interest underlying a right may be uncertain or only vaguely specified. Consider the following puzzle scholars have raised about Fourth Amendment law. Would that amendment constrain law enforcement's use of software—specifically, a type of computer "worm"—that silently scans every home computer for evidence of criminal activity?[10] To determine if this kind of massive scanning of every computer should count as a "search" subject to Fourth Amendment constraints, courts may need to know what interests the Fourth Amendment protects—and whether the mass scanning by the computer worm threatens or harms those interests.[11] Does it protect against the disruption that accompanies arbitrary police searches of a home? If so, the silent scanning of a computer user with software the owner never knows is there—until it detects something that is clearly illegal on a particular computer—doesn't seem to cause disruption. Does it

---

[10] This hypothetical was set out in Michael Adler, *Cyberspace, General Searches, and Digital Contraband: The Fourth Amendment and the Net-Wide Search*, 105 Yale L.J. 1093, 1110 (1996).

[11] *See* Lawrence Lessig, *Code and Other Laws of Cyberspace* 117 (1999).

protect against mistakes where government investigations wrongly subject innocent activity to intense and uncomfortable surveillance? If so, the software is problematic only if it erroneously identifies innocent material as evidence of a crime, a mistake the well-designed computer worm in this hypothetical scenario is assumed not to make. Or is the Fourth Amendment's purpose to protect individuals' sense that their homes are, except in extraordinary circumstances, a shelter against measures by the government to investigate and control them? If that is the underlying interest, then constant massive scanning of our computers may violate it even if it never disrupts our lives until it reliably identifies evidence of a crime.

The development of First Amendment law encounters similar forks in the road, where the right path for courts depends on what underlying interest or interests the First Amendment is supposed to shelter or advance, and the nature of the underlying interests isn't fully clear. Consider virtual reality video games the playing of which allows individuals to develop or hone certain skills (in remaining more focused when performing particular tasks, such as playing the piano, hitting a baseball, or parrying a physical attack). Or virtual reality video games that let people experience the excitement of participating in a simulated spaceship ride. Does the free speech protection that the Court has already extended to playing or video games cover such virtual training or simulation?[12] The answer may well depend on what free speech protection is for. If it is for advancing democratic discourse, many of these virtual reality experiences may have a tenuous connection to the First Amendment's core purposes—and should perhaps be more subject to regulation than is political commentary. If, on the other hand, the First Amendment's purpose is to allow individuals to exercise certain forms of autonomy—in part, perhaps by sampling experiences they can't easily obtain in the brick-and-mortar settings in which they live—then some of these virtual experiences may have just as strong a claim to staunch, generally unyielding First Amendment protection as any political statement.

---

[12] *See* Marc Jonathan Blitz, *A First Amendment for Second Life: What Virtual Worlds Mean for the Law of Video Games*, 11 Vand. J. Ent. & Tech. L. 779, 807 (2009) (explaining why the boundaries of what counts as a "video game" are unclear); Marc Jonathan Blitz, *The First Amendment, Video Games, and Virtual Reality Training*, in ed. Woodrow Barfield and Marc Jonathan Blitz, *The Law of Virtual and Augmented Reality* (Edward Elgar 2018) (asking, in part, whether and when the First Amendment doctrine has applied to video games extends to virtual reality training).

It is possible, to be sure, for courts to give force to a kind of rights protection without fully explicating the interests or underlying values it protects. In fact, as Cass Sunstein has argued, there are circumstances where it makes more sense for courts to work with an "incompletely theorized" or even "minimalist" conception of a right than a more elaborately theorized one[13] and one sense in which a court's theory might be incomplete in this sense is by working with only a vague definition of the right's underlying values or interests. Courts can often get by with a definition of the right that is broad enough to cover a range of possible interests: They can, in many cases, protect speech in a way that simultaneously protects its value in democratic discourse, in seeking truth, or as a tool individuals can use to exercise their own autonomy (or better enable their listeners to exercise theirs).[14] However, in other cases, mischaracterization of, or indecision regarding, the interest underlying a right might prevent some of the judicial analysis necessary to its development.

I have allocated significant space to considering different ways in which technological development can challenge judicial elaboration and application of constitutional rights because this discussion provides us with a framework to return to the difficult questions raised above about a right to see with technology. There are questions that arise about a right to see when we seek to bring it outside the "core" areas of jurisprudence explored in earlier chapters: where our right to see with technology is packaged with a First Amendment right to speak with it or perhaps, a First Amendment right to gather information. Or where it is a relatively uncontroversial broadening of the right we have—under other constitutional provisions—to use our own bodies and natural powers for navigating the world.

In each of these familiar areas of jurisprudence, the interests protected by a right to see with technology are the same as those protected by the right it comes packaged with. We have a right to make, share, and watch videos about matters of public interest, and perhaps also share scenes

---

[13] *See* Cass R. Sunstein, *Legal Reasoning and Political Conflict* 63, 67 (Oxford University Press 1996).

[14] *See* Thomas Emerson, *A System of Freedom of Expression* (Random House 1970); Joseph Blocher, *Free Speech and Justified True Belief*, 133 Harv. L. Rev. 439, 441 (2019) (describing theories that "accept truth as the end goal of the First Amendment" as well as others "suggesting that the First Amendment's lodestar is not truth, but democracy, personal autonomy, or some other value").

from our personal life, to advance the same interests that we further when we engage in other forms of public discourse or private conversation: our interest in becoming informed citizens and helping others to become informed citizens; our interest in learning the truth about the world; and our interest in gaining information that will also inform personal decisions of great importance to our lives. Similarly, our interest in being able to observe and learn about our immediate environment without hindrance is closely related to the other interests that are part of our personal autonomy and freedom from external interference. We don't have the minimal self-sovereignty necessary in a free society if government officials can constrain us, or prevent us from navigating or taking stock of our immediate environments, at any time and for any reason they wish to do so.

Moreover, courts have for decades faced the challenge of reconciling these interests with—or balancing them against—other crucial interests. They have balanced liberty and privacy interests we have under the First and Fourth Amendment respectively against the interests in receiving the government's help in staving off harm that can be caused by violent threats or defamatory speech, and in investigating and thwarting criminal activity.

When we seek to provide constitutional grounding for a right to see with technology that extends *outside* of this familiar jurisprudential territory, we are more likely to face uncertainty—both about the nature of the interests that underlie such a right and about what doctrine should protect these interests. I will suggest two ways we can begin to address this uncertainty when it comes to understanding what constitutional rights protect our right to observe and record our environment.

First, we might benefit from considering a few more observations about the visual system works and how the brain processes visual information. Doing so can help us understand what interests are advanced by technological enhancement of such processes, now or in future years, and how state restriction or control of how we use such technologies can harm such interests. Second, apart from these arguments for connecting a right to see (and see with technology) to interests that are familiar from the jurisprudence of First Amendment free speech law and due process law, there are also reasons to ground rights to use certain technologies of perception in *different* interests that I will link to the "right to cognitive liberty" that, since the first years of twenty-first century, has been used in a sense that is somewhat district from "freedom of thought." It encompasses a right not just to form beliefs through familiar forms of observation, and through discussion and other uses of language, but also a right to shape our minds with emerging technologies that are outside of expressive activity: Technologies including cognitive enhancement drugs and devices, neurofeedback technologies, brain-computer interface technology, and brain scanning (and other "brain reading"). Some of the interests that it is most crucial to protect as we augment and safeguard our perception belong in this category. They are interests underlying a cognitive liberty right that entails more than either freedom of speech or personal autonomy.

The Right to See and the Cognitive Construction of Our Perception

In Chapter 4, I provided a very brief overview of the biology underlying vision, drawing on the work of neuroscientists who research and write about that topic. In doing so, I expressed doubt that the specific characteristics of that biology mattered very much in answering the question of the rights we have against government interference in seeing. The protection we have against compelled surgery protects us regardless of which part of our bodies that surgery is aimed at. Similarly, being forced to take drugs that interfere with our sensory processing system would violate our personal autonomy in the same way as being forced to take drugs that undermined any other aspects of our psychological processing, or imposed other physical effects on us, such as making us weaker or more tired. A legal prohibition on being able to use our natural vision to look around us would likewise violate the same right to personal autonomy that is violated when we are prohibited from freely moving our arms or walking around an environment where we have a right to be.

However, there are certain elements of our visual processing that may make a difference in the definition of a right to see and may help legal thinkers identify when government interference in our perception raises constitutional problems. As Chris Frith emphasizes, "our brain does not simply transmit knowledge to us like a passive TV set. Our brain actively creates pictures of the world... from the very limited and imperfect signals provided by our senses."[15] As Frith notes, the nervous system has to engage in very complex processing to construct these pictures and sometimes, in doing so, creates a picture of the world that is in some respects incomplete or inaccurate. Some individuals, for example, have "blindsight." They don't consciously see certain movements or shapes even though the brain possesses, and allows them to work with and react to, information about these shapes or movements.[16] In other cases, the brain generates perceptions not from the stimuli that the eye or other sense organs receive from the environment—but through a process that can make people see what is not there.[17]

It is instructive to take note of two aspects of this complex processing discussed by neuroscientists and philosophers that raise interesting questions for this book's analysis. First, the brain does not simply build the image of the world we receive using the light from the environment as its raw material. It does so against the background of expectations about what we *should* see, given past experience. "[P]erception," Frith says, "depends upon prior belief." More specifically, "[w]hen we perceive something, we actually start on the inside: a prior belief, which is a model of the world in which there are objects in certain positions in space."[18] This "prediction" about what we should be seeing can be at odds with the sensory inputs the eye sends to the brain, in which case the brain will adjust its model. As Alva Noe says, "to a surprising extent, we see what we expect to see."[19] As Anil Seth says, whereas many people have a "bottom-up" conception of perception, whereby our brains build our image of the world from "the river of sensory data flowing into the brain,"[20] perceptions actually arise the other way around: They "come primarily from the top down, or the inside out." They arise from a "controlled hallucination" generated by "brain's perceptual best guesses" about what the world should look like, and a perceptual experience that

---

[15] Chris Frith, *Making Up the Mind: How the Brain Creates Our Mental World*, loc. 1308(Blackwell Publishing 2007).

[16] Frith, *Making Up the Mind*, *supra* note 15, at loc. 498-508.

[17] *Id.*at loc. 530.

[18] *Id. at* loc. 1965.

[19] Alva Noe, *Out of Our Heads: Why You Are Not Your Brain, and Other Lessons from the Biology of Consciousness* 135 (Hill and Wang 2009).

[20] Anil Seth, *Being You: A New Science of Consciousness* 82 (Penguin Publishing 2021).

is then adjusted in response to the visual signals entering from the world rather than wholly built from those signals. "What we actually perceive," writes Seth, "is a top-down, inside-out neuronal fantasy that is reined in by reality, not a transparent window onto whatever that reality may be."[21]

Second, the picture that is formed inside of our mind's eye may be more bare-bones than we often believe it to be. That, according to some who have studied sensory processing, is because we don't need to construct a model of the external environment inside of our heads when that environment remains just outside of us for us to consult whenever we want to focus on a particular part of it. As Alva Noe writes, "I turn my attention not to my internal model but to the world... What guarantees its availability is, first of all, its actually being here, and second, my possessing the skills needed to gain access to it. I gather the detail as I need it by turning my head or shifting my attention."[22] As Christoph Koch similarly writes, while "[f]rom the user's point of view, vision feels like an automatic process that maps external, physical reality in a straightforward manner onto the inner mental universe," that is not how perception takes place. Our mind instead "selects the few nuggets of information that are of current relevance from the vast flood of data streaming in from the sensory periphery."[23]

Perception, then, doesn't involve creating a richly detailed virtual replica inside our heads of the world outside of it but rather constant repeated interactions with that world. Since the picture we naturally receive of the external environment is far less rich in detail most of the time than many assume, those who want to use conscious perception to capture this detail—such as visual artists—have to find ways to perceive what, for most of us most of the time, is left out.[24]

These two observations about our visual system may at first seem in tension with each other. The first emphasizes the way that our perception is built from "the inside out"—with the brain playing an

---

[21] *Id.* at 88.

[22] *Noe, supra note 19.* at 140.

[23] *See* Christoph Koch, *The Quest for Consciousness: A Neurobiological Approach* (Roberts & Co; 2004), p. 153.

[24] *Id.* As Koch notes, we generally perceive the world as collections of "fairly high-level objects, such as letters on a keyboard, dogs running about, or mountains below a cobalt-blue sky" so "[i]t takes a lot of practice to sketch," as trained artists do, 'using surface patterns of differing intensity, edges, and subtle texture variations." *Id. n1, p. 153.*

active role in constructing rather than passively perceiving whatever is in the external environment. The second emphasizes the importance of frequently returning our gaze to that environment in order to extract more detail from it. Yet both of these fit together into a model of a perception which, as Seth puts it, understands it as an "intrinsically dynamic, active system, continually probing its environment and examining the consequences."[25]

These and other discussions of the visual system focus first and foremost on the way it, as well as our other sensory powers, serves an interest that is generally *not* central to discussions of First Amendment, due process, or other constitutional rights: The interest we share with other animals in surviving the challenges in our environment. As Margaret Livingstone says, to meet this interest, the pattern of neuronal activity sculpts the information we receive so it meets our needs: Whereas the initial pattern "reflects pretty accurately the pattern of light falling on the retina," later stages of neuronal activity "reflec[t] this information in an increasingly abstract, but more biologically useful, way... the higher you go in the visual hierarchy, the pattern of firing of neurons more closely approximates how you are going to interact with what is out there in the world."[26] As Anil Seth writes, "[w]e perceive the world around us in order to act effectively within it, to achieve our goals, and—in the long run—to promote our prospects of survival. We don't perceive the world as it is, we perceive it as it is useful for us to do so."[27]

What lessons do these brief observations about the biological and psychological processes underlying perception have for thinking about a right to see with technology? One is to reinforce a point that has already been made in this and earlier chapters: Our interest in perceiving effectively may not simply mirror our interest in the integrity of, and our autonomy in, our mental operations—it may be a *part* of it. Perception is not something that happens prior to cognition but is deeply interwoven with it. So if we should have a right to mental integrity or liberty, based in the Constitution or some other legal source, it should protect the

---

[25] Seth, *supra note 20*, at 115.

[26] Margaret S. Livingstone, *Vision and Art* (Updated and Expanded Edition) 24–25 (Abrams 2014).

[27] Seth, *supra note 20*, at 115-116.

mental processing we engage in as we perceive the world. Visual prostheses then can potentially consist not just of artificial eyes but also of artificial devices that replace parts of our brain responsible for certain visual processing tasks. In individuals who have a type of agnosia, for example, the eyes and ears may be functioning well, but a part of the brain responsible for sensory processing is not. Individuals with prosopagnosia, for example, cannot recognize other people by looking at them (although they can recognize them by their voices). The problem seems to stem from damage to a part of the brain called the "fusiform face area" in the inferior temporal cortex. A person who has akinetopsia can see stationary objects but cannot see motion—because of damage to the V5 visual area in the occipital lobe of the brain. In cortical blindness, a person's vision is completely or partially flawed—even though their eyes function flawlessly—because of damage to the brain's visual processing system.[28] It is possible that, in future decades, scientists will develop brain-computer interfaces (BCI) or other types of prostheses for the brain that can correct these visual deficits. A right to see, then, doesn't only entail capturing light with one's eyes. It entails a right against interference in, or disabling of, a more complex set of mental processes that lets us generate, test, and use pictures of the world. This may well entail more than the kind of knowledge creation discussed in Chapter 3, embracing even the visual inputs and components of visual processing that shape our minds outside of our conscious awareness.

There is a second tentative conclusion we can draw from the above observations about vision: Since perception generally involves continuous access to and interaction with the environment, protecting our ability to engage in it may involve protecting not simply our bodily integrity, but also our ability to have contact with the world around us. Consequently, physical presence (or perhaps telepresence) in an environment will allow us to perceive it in a way that we will probably be unable to if we have only a chance to glimpse it briefly from afar or see a two-dimensional image of it in a photograph or a frame in a video recording. Reading a verbal description of it will likely be even more removed from the experience of perceiving it directly: We may find something in that verbal description that we cannot obtain from our own viewing of the environment—namely, the perspective of and feeling experienced by the author

---

[28] See Adam Zeman, Consciousness: A User's Guide, 210–218, (Yale University Press 2002).

as *they* perceived the environment. But, in such a secondhand account, we lose the detail we have access to when we are there. Consequently, courts and legal thinkers should be hesitant to assume that the interests served by perception and intellectual exploration are satisfied even with experiences that have less perceptual detail. The videos of a place that a modern-day Louis Zemel might find on social media are weak substitutes for technology that gives him a "first-person" sense of presence there.[29]

A third lesson is that when we see or sense with technology, we might use such technology not only to *replace* lost sensory capacity (by, for example, using an artificial retina developed by engineers to replace a damaged one) or to *enhance* our senses (for example, with an artificial retina that can let see previously invisible forms of radiation)—we can *also* use it to correct or compensate some tendencies that may occasionally lead us astray even when our natural perception works the way it normally does. For example, human beings often exhibit "inattentional blindness": When something isn't expected or isn't the focus of attention, it might not be seen even when it is right in front of someone.[30] While such a failure to perceive may make sense when there is something else in our perceptual field that merits our full attention, it may also come with significant costs in certain situations. So a technological overlay on our seeing that alerts us to what we've missed might help us overcome or compensate for such inattentional blindness. So might use of recording or other image capture technology that lets us quickly review what has appeared in our perceptual field, focusing on different details in our environment on the second or third viewing than we focused on in the first. Or perhaps some augmented reality technologies that focus our attention on aspects of our environment we might normally miss. Steve Mann writes, in describing wearable computing and lifelog technology, how use of certain filters on an augmented reality device he invented "enabl[ed] [him] to see things [he] would otherwise have missed." Based on this experience,

---

[29] This point is related to yet another possible benefit of lifelogging that Steve Mann emphasized: It can "put an audience, in effect, inside the wearer's head to share a truly first person perspective." *See* Steve Mann, *Continuous Lifelong Capture of Personal Experience with EyeTap*, CARPE'04: Proceedings of the 1st ACM workshop on Continuous archival and retrieval of personal experiences (2004).

[30] Noe, *supra* note 19, at 137.

he suggested that "intelligent eyeglasses of the future" might "anticipate what is important to us" and "reveal salient details."[31] Or perhaps BCI could be developed that would make predictions of what we will see—based on past experience—less likely to make us see inaccurately.

Technologies that supplement seeing or other sensation might also counter another risk that arises in our natural perceptual processes. We can be psychologically manipulated—and not only with advanced technologies like those that involve "brain writing" to change our brain function (perhaps surreptitiously) but by others who know how to exploit humans' deep-seated tendencies to engage in biased or irrational thinking. In fact, those who design social media sites, video games, and other computer applications have been doing so in ways that many observers claim are intended to exploit our psychological tendencies in ways that make them "addictive" or elicit other behaviors from us (like purchasing a good or service) that they want to elicit. "Dark patterns," for example, are designs in social media or other applications that, according to one study, exploit "cognitive biases" in order to push users to "purchase goods and services that they do not want or to reveal personal information they would prefer not to disclose."[32] Like other mental processing, our visual processing can also be a target of manipulation. Because, as Noe says, "to a surprising extent, we see what we expect to see... we are very suggestible."[33] This can take a very innocent form: When we attend a magic show, we typically *want* to be deceived—and impressed by the magician's seemingly impossible trick. As viewers of art, we may enjoy art that is created by an artist who is aware of how our minds often perceive colors and shapes and uses this knowledge of our psychological processes to create art that has a powerful effect on us. But such deception can also take a harmful form and so we may welcome tools or other technologies that allow us to alter the way we see the world to minimize the chance we will be victims of such manipulation. Technological augmentation of technology can thus play a role in a kind of perceptual self-defense system. Wearable computers, for example, might be used to

---

[31] Steve Mann, *Wearable Computing: A First Step Toward Personal Imaging*, Computer 30(2), Feb. 1997: 25–37, 28.

[32] Jamie Luguri and Lior Jacob Strahilavitz, *Shining a Light on Dark Patterns*, 13 J. Legal Analysis 43 (2021).

[33] Noe, *Out of Our Minds, supra* note 19, at 137.

filter out advertisements or combat technological interfaces designed to exploit psychological weaknesses.[34]

To some extent, lifelogging and ubiquitous recording aimed at those who may surveil or manipulate the recorded footage is an instance of this. As I have already noted in Chapter 1, when citizens record those who surveil them, they engage in what Steve Mann, Jason Nolan, and Barry Wellman call "sousveillance"—a "watching of watchers" that can "hol[d] up a mirror to society" and challenge the asymmetry between powerholders and those who they surveil with massive camera systems.[35] Legal scholars such as Jocelyn Simonson, Scott Skinner-Thompson, and Timothy Zick have explored how constitutional law and other law can enable sousveillance in the form of citizen creation of video records—and also stress how the openness of citizen recording also contributes to democracy and expression. Skinner-Thompson writes, "Open and visible sousveillance of government actors can begin to exert influence on their behavior as the behavior is unfolding, and it also serves as an expression of critique of that behavior."[36] Simonson notes that sousveillance can provide a form of deterrence against abuse, functioning as "a technique of deterrence" of government abuse.[37] Zick considers how recording has provided such deterrence at protests.[38] Mann also describes how sousveillance enabled by lifelogging technology can counter another harm of surveillance, which is the possibility that surveillance recordings by officials or other external actors will be "taken out of context."[39] As noted below, Bobby Chesney and Danielle Citron have recently suggested that

---

[34] See Steve Mann and Woodrow Barfield, International Journal of Human–Computer Interaction, 15(2), 205–208 (2015) (noting that "mediated reality may [ ] serve as a framework for filtering out real world (advertising billboards; Mann, supra note 29, at 2 (explaining that "EyeTap technology has been used to remove the annoying visual propoganda that plagues our urban environments.").

[35] See Steve Mann, Jason Nolan, and Barry Wellman, *Sousveillance: Inventing and Using Wearable Computing Devices for Data Collection in Surveillance Environments*, Surveillance & Society 1(3): 331–355 (2002).

[36] Scott Skinner-Thompson, *Recording as Heckling*, 108 Geo. L.J. 125, 139 (2019).

[37] Jocelyn Simonson, *Copwatching*, 104 Cal. L. Rev. 391, 416 (2016).

[38] Timothy Zick, *Clouds, Cameras, and Computers: The First Amendment and Networked Public Places*, 59 Fla. L. Rev. 1, 20 (2007).

[39] See Mann, supra note 29, at 2.

pervasive recording of one's own life can similarly provide a defense against manipulation or defamation by deepfake creators.[40]

Fourth and finally, it is important—in thinking about rights to see or sense—to keep in mind that the interest that students of visual processing often focus on, the interest in being able to interact with our environment in a way that increases our chances of survival is *not* the interest that is most threatened by many forms of worrisome government interference in seeing. Consider, for example, John Humbach's explanation for why the First Amendment should be understood to include a "right to observe" the world: It is, at least in part, that without the chance to view the world around us, we are unable to engage in the kind of critical reflection about politics that democracy requires: "What better way to squelch debate about some government abuse," he asks, "than by forbidding people to observe it, even when it occurs in plain sight?"[41] However, a visual and sensory system that has evolved to let us survive in the physical environment may not necessarily be as well designed to further our interest in observing what citizens need to learn about to monitor and reflect upon their government's policies. The latter may well require being able to observe events that occur far from their immediate environment—with access to others' recordings, global mapping programs (like Google Earth), or extended reality technology that can let us immerse ourselves in different settings. The effectiveness of democracy may also be safeguarded by technologies that enable citizens to perceive certain aspects of modern life our visual systems have not evolved to perceive: As Mann writes, technology may not only enable us to engage in "sousveillance" but also "metaveillance," whereby we can perceive the presence of surveillance technology even when is hidden from our natural sight.[42] The same may be true of other core interests that the First Amendment and other rights are often described as safeguarding. Many analyses of First Amendment rights, for example, emphasize our interest in autonomy, our

---

[40] *See* note 86–87 *infra* and accompanying text.

[41] John A. Humbach, *Privacy and the Right of Free Expression*, 11 First Amend. L. Rev. 16, 43–44 (2012).

[42] *See* Steve Mann, *Surveillance (Oversight), Sousveillance (Undersight), and Metaveillance (Seeing Sight Itself)*, 2016 IEEE Conference on Computer Vision and Pattern Recognition Workshops (CVPRW) at 1408–1417.

interest in protecting "the development of [our] rational capacities."⁴³ But our autonomy may sometimes be best served by capturing and evaluating much more extensive information about our day-to-day lives than the amount of information we normally extract from our environment. Lifelogging technology may thus help to further our interest in autonomy more effectively than does our natural visual processing. In short, technology that supplements our perception may be necessary to best fit it to interests that make it crucial for us to have a right to perceive.

Cognitive Liberty, Remote Sensing, and the Constitution

There is another set of interests that make it important for us to have a right against government interference in the way we perceive the world—and those are the interests that underlie our right to cognitive liberty. The term "cognitive liberty" was coined by Richard Glen Boire and Wren Sententia, who founded a center and journal dedicated to the protection of our right to cognitive liberty. Sententia has said the term "updates notions of 'freedom of thought' for the twenty-first century by taking into account the power we now have, and increasingly will have, to monitor and manipulate cognitive function."⁴⁴ Boire has similarly described it as a "right to control one's own consciousness"—not only by reading books, writing essays, and taking other actions already clearly protected by the freedom of speech, but also by using drugs or other means to alter one's thought processes and feelings in particular ways.⁴⁵ In recent years, other authors have similarly used the term "cognitive liberty" to denote something distinct from a traditional reference to a right to "freedom of thought." Nita Farahany, for example, uses the term "cognitive liberty" to refer to "the right to self-determination over our brains and mental experiences." This, she says, is broader than a

---

⁴³ Christina E. Wells, *Reinvigorating Autonomy: Freedom and Responsibility in the Supreme Court's First Amendment Jurisprudence*, 32 Harv. C.R.-C.L. L. Rev. 159, 170 (1997).

⁴⁴ Wrye Sententia, *Neuroethical Considerations: Cognitive Liberty and Converging Technologies for Improving Human Cognition*, Annals of the New York Academy of Sciences. 1013 (1) (2004).

⁴⁵ Richard Glen Boire, *On Cognitive Liberty*, Part I, 1–3 Journal of Cognitive Liberties 1(1)(2000).

right to freedom of thought, which is only one component of it, since it also includes rights to "mental privacy" and "self-determination."[46] Other writers have continued to use others' terms, such as "freedom of thought" or "mental self-determination" to encompass activities one takes to reshape one's brain function or mental processes.[47]

How then might rights of perception be conceived as a part of such a right to cognitive liberty? In short, technologically altering what we see might not only be crucial to knowledge creation. It may be a crucial component of how we shape what we think and feel. The simplest case is presented not by video recordings or other technologies for perceiving actual events but rather by virtual reality technology. The exploration of fictional virtual environments and the perception of fictional virtual events made possible by VR is a kind of technological aid to our power of imagination. When we use VR simply to explore a fantastical landscape of our own creation, or to explore—in more vivid form—something that someone else has imagined, we are doing something very like immersing ourselves in a dream or daydream. But instead of conjuring up a picture with the power our brain gives us to privately visualize something in our mind's eye, we are instead creating sensations that can generate a perception this event from the outside.

Consider, the virtual reality most readily available to consumers of the early twenty-first century. It generates an illusory environment with light generated inside of a "head-mounted display" (or HMD) that sits over a person's eyes—according to a program run inside of an attached computer. Sound is fed to the user through headphones.[48] And in some VR systems, individuals also receive tactile sensations through gloves or even bodysuits.[49] Prototypes for these kinds of computer VR systems were developed in the 1960s, by engineers such as Ivan

---

[46] Nita A. Farahany, *The Battle for Your Brain: Defending the Right to Think Freely in the Age of Neurotechnology*, 214 (St. Martin's Press 2023).

[47] For a detailed recent analysis of the history and meaning of the term "cognitive liberty" and its relationship to discussion of freedom of thought, see Jan Christoph Bublitz, *A History of Cognitive Liberty*, in Jan Christoph Bublitz and Marc Jonathan Blitz, *The Law and Ethics of Freedom of Thought*, Volume 2: Cognitive Liberty, Mental Privacy, and International Law (forthcoming Palgrave Macmillan 2025).

[48] *See* Michael Heim, 7 (Oxford University Press 1998).

[49] *See* Frank Biocca and Ben Delaney, *Immersive Virtual Reality Technology*, in Frank Biocca and Mark R. Levy *Communication in the Age of Virtual Reality* 3, 5 (Routledge 1995).

Sutherland and Thomas Furness.[50] Today, individuals, play video games or immerse themselves in various environments (fictional or real) with systems produced by Oculus (a company owned by Google), by HTC, or by Apple. Other systems use alternative means to immerse individuals in an illusory environment. For example, this is true of VR systems created by the artist and engineer Myron Krueger as well as VR systems modeled on the "Cave Automatic Virtual Environment" (or "CAVE") pioneered by researchers at the University of Illinois.[51] In that kind of system, what we see is projected onto walls that surround the individual rather than an HMD we wear over their eyes. The user experience is often supplemented with "stereo shutter glasses" that can provide users with the sense that the images they see on the walls are three-dimensional rather than flat, and with equipment that tracks users' head movements and eyes and responds in a way that mimics their interactions with actual environments.[52]

Since the experience produced by such VR environments is a kind of externally generated dream sequence, it is not hard to see why one might consider it to be an example of the kind of conduct that should be shielded by a right to freedom of thought or cognitive liberty. The state would certainly violate such a right if it found a way to interfere in the dreaming or daydreaming we engage in by closing our eyes and conjuring mental imagery. Why should it not also be understood to do so when it bars us from artificially creating an identical or similar private mental experience? To see, in this case, is the functional equivalent of imagining.

The case for treating many uses of VR as exercises of a right to freedom of thought or cognitive liberty is bolstered by the uses that many individuals make of this technology. It has become an important tool of psychotherapy. Many individuals, for example, seek to conquer phobias

---

[50] Ivan E. Sutherland, *The Ultimate Display*, 2 Proceedings of IFIP Congress 506–08 (1965), republished in Bruce Sterling, *Augmented Reality: "The Ultimate Display""by Ivan Sutherland, 1965*, Wired, Sep. 28, 2009, available at https://www.wired.com/2009/09/augmented-reality-the-ultimate-display-by-ivan-sutherland-1965/; *See* Fred Moody, *The Visionary Position: The Inside Story of the Digital Dreamers Who Are Making Virtual Reality a Reality* xxiii (1999).

[51] See Myron W. Krueger, *Artificial Reality* II xiii (1991); Carolina Cruz-Neira, Daniel J. Sandin, Thomas A. DeFanti, Robert V. Kenyon and John C Hart *The CAVE: Audio Visual Experience Automatic Virtual Environment*. Commun. ACM. 35 (6): 64–72 (Jun. 1 1992).

[52] John Vince, *Virtual Reality Systems* 14 (Pearson Education 1995).

by confronting them in virtual environments (where they can face their fears repeatedly and monitor and adjust their reactions).[53] It also has been used to supplement or replace the visualization that many individuals use to mentally rehearse a particular task before they engage in it and prepare to meet challenges with calmness and concentration. An artificially generated VR experience should be just as shielded from state intrusion as a private visual experience we generate with our natural power of visualization. In fact, one might even argue that there is no need to draw on a new right to cognitive liberty here: Creating a painting, animation, or video game is, in the twenty-first century, understood to be an act of artistic expression protected as a kind of "speech creation." Viewing such a painting or animation or playing such a video game is also protected—because the First Amendment allows us to serve as the audience for each others' speech, or, for that matter, the art or other expression we create ourselves.[54] A virtual reality experience, one might therefore argue, would have the same First Amendment status: Even if it looks and feels like a real experience, it would be considered an artistic creation.

COGNITIVE LIBERTY, REMOTE SEEING AND SENSING, AND COMPREHENSIVE RECORDING

A more complex challenge arises when what we claim a right to see and explore is not an imaginary landscape or a sequence of fictions—but rather a real place or event. First, what we see cannot be as private in the same way. When Louis Zemel wished to see Cuba in 1962, he could not do so by staying in his own home or some other private space within the United States and simply immersing himself in an exercise of his imagination or a lucid dream. He needed to be in a place where he could directly observe the country he wished to see—or he needed somebody to do it for him and share a visual record of their experience. In order for someone to capture the light from a real city, or the light that reveals if a company is complying with environmental laws or laws against the abuse of animals, they have *to be where* the light can reach their camera. They may, in attempting to do so, find their exercise of seeing blocked

---

[53] *See* Kate E. Bloch, *Virtual Reality: Prospective Catalyst for Restorative Justice*, 58 Am. Crim. L. Rev. 285, 313 (2021) (discussing use of "virtual reality scenarios also furnish an environment for forensic mental health work").

[54] *Sorrell v. IMS Health Inc.*, 564 U.S. 552 (2011).

by property rights. They might also find it countered by privacy rights: Their interest in freely seeing their environment, the state may argue, is at odds with someone else's powerful interest in *not* being seen.

More generally, while perceiving a particular environment is, of course, an exercise of one's mental power, it is much more than that. It requires not just an individual's perception but also the presence in a particular physical environment—either of the individual or of someone capturing a record for them. The individual has a kind of sovereignty over their own perceptual process but may not have it over the environment to be perceived.

What then does such a perception of an external environment have to do with cognitive liberty or freedom of thought? It is helpful to start by recalling that, where video recording counts as speech creation (as it often does), it cannot simply be evicted from the realm of First Amendment protection on the ground that there are certain places where individuals do not have a right to locate themselves. That doesn't mean that there aren't limits on rights to record. Federal courts that have addressed this question have almost all recognize that individuals at the very least have a right to record public officials' actions in public places.[55] Some have also described the right as a broader right to make recordings of all types of "matters of public concern" in public places.[56]

And some have extended the same right—to record "matters of public concern"—even to privately-owned environments where a person has a right to be. For example, multiple courts have now found that animal rights activists who recorded the apparent mistreatment of animals in agricultural facilities after applying for, and receiving, employment there had a right to make such recordings.[57] Such holdings do *not* clearly indicate that individuals have any First Amendment right to record matters of purely private concern in public space, let alone in other people's private

---

[55] *See* ACLU of Ill. v. Alvarez, 679 F.3d 583, 595 (7th Cir. 2012); Irizarry v. Yehia, 38 F.4th 1282, 1289 (10th Cir. 2022); Fields v. City of Philadelphia, 862 F.3d 353, 360 (3d Cir. 2017); Turner v. Lieutenant Driver, 848 F.3d 678, 690 (5th Cir. 2017); Glik v. Cunniffe, 655 F.3d 78, 82 (1st Cir. 2011); Smith v. City of Cumming, 212 F.3d 1332, 1333 (11th Cir. 2000); Fordyce v. City of Seattle, 55 F.3d 436 (9th Cir. 1995).

[56] *See* Fordyce v. City of Seattle, 55 F.3d 436, 439 (9th Cir. 1995) (speaking of a "right to film matters of public interest"); Askins v. U.S. Dep't of Homeland Sec., 899 F.3d 1035, 1043 (9th Cir. 2018) (speaking of the same right).

[57] Alan K. Chen and Justice Marceau, *Truth and Transparency: Undercover Investigations: In Twenty-First Century* 162, 165 (Cambridge University Press 2023).

homes or businesses, except perhaps when they are recording their *own* private activities in their own homes or other places where they have a right to be.[58] Moreover, even where individuals do *not* have a right to be, this does not mean they can be prevented from making a recording on a matter of public concern if the government's aim in stopping them is to prevent them from sharing, or the public from learning about, a particular matter of public interest that is not shielded by any privacy right.

All of this can be justified by the interests that are often taken to be at the heart of First Amendment free speech protections. First, as the Court has often stressed, it is not any and all expression of our perceptions that courts identify as having First Amendment value. It is rather "core political speech" or, as the Supreme Court has described it, speech that presents "ideas for the bringing about of political and social changes desired by the people."[59] Consequently, it is not surprising that courts have suggested that protection of recording might focus first and foremost on recording of these types of subjects.

Moreover, as Alan Chen and Justin Marceau point out, it should also not be surprising that this protection for recording matters of public concern extends into some private environments—like agricultural facilities where animals are bred and raised for food.[60] A significant amount of information that is crucial for the public to know about—including deeply harmful violations of laws on the environment, animal protection, discrimination, and other matters—can only be revealed by determined investigative reporting that draws on information sources on the property of private businesses.[61] Video recording can often reveal such harms more effectively than written descriptions. So it is not surprising that courts have extended First Amendment protection to this type of recording of matters of public concern. Moreover, Chen and Marceau observe, many courts have stressed, in recognizing a First Amendment right to record, that recording is "speech creation" and that logic applies as much to recording in private spaces as to recording in public spaces.

---

[58] Id.

[59] Roth v. United States, 354 U.S. 476 (1957) ("speech and press was fashioned to assure unfettered interchange of ideas for the bringing about of political and social changes desired by the people").

[60] Chen and Marceau, *supra* note 51.

[61] *Id.*

This kind of framework hardly resolves all of the questions that may arise about when recording that counts as speech creation is protected by the First Amendment and when it is not. In the first place, while the First Amendment certainly protects discourse on matters of public concern, its value to individuals goes much further. As I have argued in the past, many individuals who explore Google Streetview or other databases of recorded images may be far more interested in matters of personal rather than public interest:

"An individual might use Google or Bing Maps to revisit a neighborhood from a previous era of her life or some other 'old haunt' of little significance to most other members of the public. Or instead of visiting scenery from the past, she may wish to scout a possible home for the future. She may wish to get a sense of what it feels like to look at the buildings, or foliage, of a town they are thinking of moving to, or going to school in. Or to immerse herself visually in a far-away locale that happens to provide the historical setting for an intriguing book that [s]he has just read."[62]

It seems arbitrary, and at odds with First Amendment values, for First Amendment law to insist to these individuals that they are focusing on the wrong passions or interests.

Moreover, where someone is located—whether they are in a public or private space—is not the only important variable that courts must consider. Recording can, as we have seen throughout this book, be supplemented with numerous technologies. Pervasive automated recording may threaten privacy much more than a single person with a camera.[63] Recording from drones or other aerial vantage points may do so as well. A camera on a drone or helicopter might capture footage from a home's backyard, porch, or interior area adjacent to a window—especially with the aid of magnification.[64] Even if it is flying over public space in a public airway, its zoom-equipped cameras might be pointed

---

[62] Marc Jonathan Blitz, *The Right to Map (and Avoid Being Mapped): Reconceiving First Amendment Protection for Information-Gathering in the Age of Google Earth*, 14 Colum. Sci. & Tech. L. Rev. 115 (2013).

[63] *Id.*

[64] *See* Margot Kaminski, *Privacy and the Right to Record*, 97 B.U. L. Rev. 167, 170–171 (2017).

at private property.⁶⁵ Even if courts hold that cameras, under the First Amendment's shielding, may capture what is visible to the public in public spaces, this doesn't make clear what types of magnification or other enhancements are included within that right.

It also doesn't fully address other types of seeing with technology. As the preceding chapters have emphasized, many of the ways we enhance our perception not only don't involve communication—they don't necessarily involve the creation of a record that can be communicated. Live video streams, for example, might leave a record of what they show only in the mind of the viewer.

Moreover, even where our perceptions are recorded, we mischaracterize the value they have for individuals if we emphasize only the potential role that a recording has in communication. Just as a private diary can provide great value to its author even if it is not shared nor ever meant to be shared, so a privately viewed recording or XR experience can have value for the person as they view it themselves. It can allow for greater self-understanding, development of skills or interests, and intellectual exploration even when such perception is not a part of speech. When video or other records or technological enhancements of perception are used in this way, we can understand them as tools of cognitive liberty: They allow individuals to shape the way their minds work. They clearly do so when they allow us to give form—in virtual or other extended reality—to our imagination. Or allow us to interact, and train our minds with, a scene that is designed by someone else for that purpose. But they also do so when they capture and allow us to perceive images of events in the real world.⁶⁶

Consider lifelogging again. Not all lifelogging takes the form of video recording. Lifeloggers use all kinds of different data to construct moment-by-moment archives of their existence, or "quantify" their histories: GPS records of movement through physical space,⁶⁷ digital trails of their

---

⁶⁵ Marc Jonathan Blitz et al., *Regulating Drones Under the First and Fourth Amendments*, 57 Wm. & Mary L. Rev. 49 (2015).

⁶⁶ *See* Shiqi Jiang, et al., *Memento: An Emotion-Driven Lifelogging System with Wearables*, 15 ACM Trans. Sen. Netw. 1 (Jan. 2019).

⁶⁷ *See* Go Tanaka, et al., *GPS-Based Daily Context Recognition for Lifelog Generation Using Smartphone*, 6 Int'l J. of Adv. Comp. Sci. and Appl. 2 (2015).

movement through cyberspace,[68] data from health and fitness trackers,[69] and brain-computer interface data that registers emotional states. Video recording allows lifeloggers to access a more vivid record of each event they have experienced. As Anita Allen notes in an illuminating discussion of the implications of lifelogging for privacy and privacy law, computer devices might soon "enable people fully and continuously to document their entire lives."[70] As Allen points out, this technology could act as a powerful memory enhancement device. It can be "a less fallible and selective adjunct to human memory."[71] As noted in Chapter 1, this was one of the benefits that Mann emphasized in introducing the technology for lifelogging and demonstrating it: Lifelogging can serve as a "visual memory prosthetic" of unprecedented power. Having such a comprehensive video archive can, of course, enable to individuals preserve and revisit significant memories, and also to analyze themselves and change their behavior more effectively. In fact, science fiction writers imagined, many decades ago, how a viewable catalog of a person's memories might enable therapists to work with their clients to understand and address aspects of their lives they wish to change.[72]

Even video records—or remote perception—of a kind that gives people access to experiences far beyond their own personal experience can contribute to individual autonomy in a similar way. This can include the chance to navigate through far-away locations on Google Earth or other mapping programs. It can also include receiving remote video of those locations, or perhaps, the chance to explore them (or perceive a sequence of events that occurs in that location) with technology that gives a person telepresence there—that is, the chance to insert themselves into an immersive three-dimensional copy of the scene they would encounter—to see, hear, and perhaps be able to touch the surroundings in which they have telepresence as though they were physically there. Being

---

[68] Nancy Messieh, *Keeping a Lifelog: The Definitive Guide*, TNW, Jul. 11, 2011 (describing logging of online and computer activity).

[69] *Id.* (describing use of fitness trackers for lifelogging).

[70] *Id.*

[71] *Id.* at 50.

[72] See Philip K. Dick, *The Man Who Japed* (First Mariner 2012 [1956]) (imagining a therapt session where psychotherapists are not limited to asking their patients to recount their childhood memories: They can use "drugs and gadgetry" to view them).

able to virtually visit a remote location in this way of course allows individuals to do what Louis Zemel wished to do but without physical travel: To understand a particular place as part of an effort to be an informed citizen. However, such technology also helps promote not just individuals' participation in democratic discourse but also their autonomy and self-development. In his classic 1859 essay, *On Liberty*, the philosopher, John Stuart Mill emphasized that in order to develop the power to shape their own intellects and choose (or build) a way of life that fits their personal nature, individuals need more than "freedom" in the sense of being free of unwarranted interference by the state or society. They also need to encounter "a variety of situations." It is the encounter with this variety, he said, that gives rise to "individual vigour and manifold diversity."[73] One of the hallmarks of individual freedom in modern pluralistic societies is that it supplies individuals with a diversity of ideas: In the twenty-first century, they find it not only in their own stock of personal experiences, but also in the much wider range of recorded experience in libraries and countless blogs, web pages, and essays on social media. These written accounts of others' experiences, while valuable, give individuals a second-hand account of what others have seen and heard, and their thoughts about those experiences. It is not surprising that individuals find it invaluable to encounter many distant locations (and the cultures of those places) directly—to see them for themselves. And technologies that provide a way to explore a "library" of other experiences and places from their computer, perhaps exploring innumerable destinations in a single day. As I noted in past work, the inventor, Oliver Wendell Holmes, Sr. wrote with enthusiasm that a library of immersive images viewable on a stereoscope would enable any person to visually explore any place—or "any object, natural or artificial"—in the same way as they might explore fictional environments in the stories they find in a library.[74]

In short, the interests that underlie our use of these perception-enhancement technologies are not simply the interests that justify a right to free speech—and to freely speak with the aid of, and through, technology. They are also interests that are better understood as underlying an emerging right to cognitive liberty—or, a right one might conceive of as a

---

[73] John Stuart Mill, *On Liberty and Other Essays* 17, 37, in ed., John Gray (Oxford University Press 1991).

[74] Blitz, *The Right to Map*, supra note 56.

twenty-first-century incarnation of freedom of thought. This right should entail a right both to produce thought-equivalents with technology—by, for example, enabling one to give vivid form to one's thoughts or reshape our thoughts and feelings with technology—and, I have argued, to reshape the self that does the thinking.[75]

Privacy Threats and Deepfake Distortions

Apart from asking how these technologies might further our mental autonomy or promote other interests that can ground a right to freedom of thought or "cognitive liberty," legal thinkers and courts should also, as I noted in the previous chapter, look at what is often on the other side of the judicial analysis: The harm or risks of harm that the exercise of the conduct purportedly covered by a right might generate for other significant interests individuals have. When we are talking about technologies of perception, the most significant of these are interests in privacy (or more general dignitary interests), in undistorted perception, and in avoiding certain ways that new technologies for mediated seeing arguably worsen our lives by, for example, leading us to respond to certain challenges with escapism. Some have warned that virtual reality may be addictive for us and weaken our connection to local communities—something the rise of smartphones and social media has, according to some observers, already done. As I will note in this section, the threat that a certain right (or extension of a right) may raise to such interests might be viewed as a basis for what I will call "counter-coverage"—that is, an argument that the right shouldn't have a scope or "coverage" that will block the state from taking measures to protect privacy or other key interests. So, for example, if recognizing a right to make continuous recordings of our environment would make it impossible for people in a park or at an outdoor seating of a café to have private conversations without worrying that their conversations were being captured, this could be treated as a reason for defining the "coverage" of the right more narrowly—so it doesn't prevent restriction of such technology.

It is useful to briefly review two of the most significant potential harms raised by recording and by other technologies of extended perception:

---

[75] See Marc Jonathan Blitz, *Freedom of Thought for the Extended Mind: Cognitive Enhancement and the Constitution*, 2010 Wis. L. Rev. 1049 (2010)

the threat to privacy and the threat of fabricated and distorted perception. The threat to privacy raised by emerging surveillance technologies is a familiar one in Fourth Amendment law. In a world covered by cameras, it is increasingly impossible to escape the government's gaze. This is particularly true where the cameras not only line streets (and can be mounted on drones) but are equipped with magnification capacities and often with facial recognition software or other technology that can analyze and extract information from capturing images or sounds. Fourth Amendment jurisprudence thus illustrates how augmenting the way observers see and sense can erode individual privacy—and while the focus of the Fourth Amendment is government surveillance, it can provide some lessons for how we think about protecting privacy if the First Amendment or other constitutional provisions extend to observations by individuals.

In fact, although this book began by focusing on the First Amendment, it is the jurisprudence of the Fourth Amendment's protection against "unreasonable searches and seizures," that has more to say about seeing, both with unaided vision and with technology. English and American courts for centuries have been interested in the relationship between seeing and individual security against government searches.

Those cases have emphasized that the Fourth Amendment's protection against unreasonable searches protects against "a too permeating police surveillance."[76] But these cases also have provided other interesting guideposts about what types of observations by police—and with what technology—are and aren't too permeating. One thing they have repeatedly stressed is that police do not offend this right—they do not engage in any kind of constitutionally-restricted "search" let alone an "unreasonable" one—when they do nothing more than visually scan their surroundings. This point is central to Fourth Amendment law—and was made even before the Fourth Amendment came into being. In 1765, Lord Camden wrote an often-cited opinion in Entick v. Carrington, an English case that inspired the Framers of the U.S. Constitution to enact the Fourth Amendment. "[T]he eye cannot by the laws of England be guilty of a trespass," said Camden. The Supreme Court has cited this in recent years to emphasize that "[v]isual surveillance" by police—when it occurs without any other intensive investigation of the home—is

---

[76] *Carpenter v. United States*, 585 U.S. 296, 305 (quoting *United States v. Di Re*, 332 U.S. 581, 595 (1948)).

"unquestionably lawful."[77] In other cases, the Supreme Court has emphasized that—while we are protected against government surveillance that intrudes into our reasonable expectations of privacy—we can't claim we've been subject to such intrusion when police do nothing more than look at us, or look at our homes, from a public street. "The police," the court said on another occasion, "cannot reasonably be expected to avert their eyes from evidence of criminal activity that could have been observed by any member of the public."[78] To merely observe what is left open to observation, in other words, is within the leeway that the Constitution gives police to conduct investigations free of Fourth Amendment limits.

Nor do police lose any of this freedom to vigorously gather evidence from the environment when they supplement their vision or other sense in ways that aid their investigation of crime but do not radically reduce citizens' privacy. In *Knotts v. United States*, for example, police hid a radio transmitter in a canister of methamphetamine ingredient purchased by a suspect so they could track his car even if they lost sight of it. The Court said this technological tracking did not trigger Fourth Amendment protections: "Nothing in the Fourth Amendment," it said, "prohibited the police from augmenting the sensory faculties bestowed upon them at birth with such enhancement as science and technology afforded them in this case."[79] It had a similar reaction in *Dow Chemical v. United States*, where Environmental Protection Agency (EPA) officials wanted to do a follow-up investigation of a chemical plant and Dow Chemical, which owned the plant, refused its consent for EPA officials to enter without a warrant from a judge. The EPA responded by flying a plane over the plant and essentially conducting their investigation from the air: They hired a commercial aircraft and took photographs of the plant with a camera that could powerfully magnify details of the plant, allowing "identification of objects such as wires as small as ½-inch in diameter."[80] Despite the added power the camera and airplane gave the government to scrutinize Dow Chemical's property, the Court found no Fourth Amendment concern:

---

[77] Kyllo v. United States, 533 U.S. 27, 31 (2001).
[78] California v. Greenwood, 486 U.S. 35, 41 (1988).
[79] United States v. Knotts, 460 U.S. 276, 282 (1983).
[80] Dow Chem. Co. v. United States, 476 U.S. 227, 238–239 (1986).

"The mere fact that human vision is enhanced somewhat, at least to the degree here," it said "does not give rise to constitutional problems."[81]

This doesn't mean there is never a constitutional problem with law enforcement's use of technology to extend law enforcement's powers of perception. In *Dow Chemical* itself, the Court stressed that, unlike a camera with magnification capacity, an "electronic device to penetrate walls or windows so as to hear and record confidential discussions of chemical formulae or other trade secrets would raise very different and far more serious questions."[82] As I mentioned before, the Court found the Fourth Amendment was violated when law enforcement used a thermal imaging device to gather information from the interior of a home from a public street.[83] And even when police collect information about a person's movements in public spaces—where that person can be observed by passersby—that doesn't mean police can, without satisfying the Fourth Amendment's requirements, create (or obtain from a cell phone carrier) a comprehensive record of such movements. As the Court said in 2018 in *Carpenter v. United States*, even if a brief recording of a person by a camera overlooking a street is not a search, it is a Fourth Amendment search for police to compile "a detailed chronicle of a person's physical presence" in particular places "compiled every day, every moment, over several years."[84] In short, while the Constitution does not restrict government surveillance that is "somewhat" technologically enhanced, certain enhancements raise too significant a threat to private to remain wholly unrestricted. Court cases on how Carpenter applies to video surveillance haven't all agreed. However, in one subsequent case, the Court of Appeals for the Fourth Circuit found that the city of Baltimore acted in violation of the Fourth Amendment when it hired a company to make round-the-clock recordings of streets from an aerial vantage point to better enable to it to battle crime and officials were able to access those records without obtaining a warrant.[85]

---

[81] *Id.*

[82] *Id.*

[83] Kyllo v. United States, 533 U.S. at 37.

[84] Carpenter v. United States, 585 U.S. 296, 315 (2018).

[85] Leaders of a Beautiful Struggle v. Baltimore Police Dep't, 2 F.4th 330, 343 (4th Cir. 2021).

What does this mean for thinking about a right to see or sense with technology? In the first place, it emphasizes that as the means by which we perceive the world increasingly allow for extensive remote sensing and increasingly sophisticated and extensive storage of what we record, the more they threaten our privacy interests. Even as the enhancement of perception one finds in lifelogging or opportunities for remote seeing or sensing enhance some of the interests that underlie cognitive liberty, they threaten privacy interests that, as noted in Chapter 2, are equally crucial for autonomy.

Earlier in the chapter, I noted some reasons why courts would think it *necessary* to expand a right to ensure it covers new, more technologically-sophisticated or-enhanced versions of the conduct it already covers. Courts want to make sure that the First Amendment protects speech in the form it occurs in the twenty-first century—in social media posts and websites. They don't want it to be rendered ineffective or irrelevant by changes in technology. They similarly want to make sure that the Fourth Amendment protects against modern forms of electronic surveillance not only against the kind of physical searches that were possible when the Constitution was enacted. Advances in technology may also require courts to recognize other adjustments to the scope of a right in order to ensure that it advances the underlying interest it is supposed to safeguard. At times then, technological change requires expanding the coverage of a right.

However, technological change may also justify what I will here call "counter-coverage": It may weigh *against* extending a right's coverage where doing so would prevent us, or our representatives in government, from safeguarding interests against uses of technology that would, if shielded by such a right, present a greater threat. Where advances in camera technology erode our privacy, we may need legal protection to restore it—but where use of such camera technology is insulated against all government interference by a right, then it is insulated against measures that can keep non-government actors from invading privacy. This is therefore a reason to find the right is limited enough to provide the government leeway to protect against the technologies of surveillance similar to those investigative uses of technologies that courts have found require a warrant (or some other constitutional safeguard) when employed by the government.

How might a right to see with technology be justified in the face of such an argument for counter-coverage? One possible response is to argue that there is an alternative to abandoning such a right even where it creates risks to privacy. We should ask whether a right to augment our perception with technology can coexist with sufficient room for a government to address the threats that might accompany such augmentation. Rather than viewing a right as inevitably shutting down *all* regulation in the sphere it covers—in order to prevent the government from making *any* entry into a zone *entirely reserved* for individual autonomy or some interest that must generally remain untouched by the government— courts sometimes view rights as playing a more nuanced role. They assure that certain interests, in privacy or free expression, are given sufficient weight even as they are balanced with, or reconciled to, a social order that in some respects demands their limitation. There is little dispute governments must sometimes subject political demonstrators to rules in order to protect the rights of others using the same streets—but that doesn't mean that the demonstrators are left with no right at all to freedom of expression.[86] Similarly, individuals' privacy rights must sometimes be balanced against the need for crime prevention (and security more generally) especially outside the home—but that doesn't mean the privacy right simply vanishes: Even in airports or regulated workplaces, for example, certain types of government searches would violate Fourth Amendment privacy rights.[87] The same then might be true, as noted in Chapter 3, of a right to observe the world and enhance what we see (or even build what we see from scratch) with technology. It is possible government can combat any threats it poses to privacy or of manipulating our senses and do so while respecting certain limits still imposed by a non-absolute form of the right. At times, perhaps this less powerful version of a right to see can leave the government with the room it needs to protect the public. In other contexts, it may be the case that any such right would leave us too vulnerable to privacy invasions or manipulation of our thought. In fact, in cases like this, it may well be the case that the interests that underlie a right to see are, in a sense, on the other side of the equation: They

---

[86] *See* Porter v. Gore, 354 F. Supp. 3d 1162, 1171 (S.D. Cal. 2018) ("content-neutral restrictions are subject to intermediate scrutiny and 'must be justified without reference to the protected speech's content,' and 'narrowly tailored to serve a significant government interest, leaving open ample alternative channels of expression'").

[87] Camara v. Mun. Court of San Francisco, 387 U.S. 523, 535 (1967); New York v. Burger, 482 U.S. 691, 703–04 (1987).

demand restriction of companies and other private actors rather than of the government.

How then, one might then ask, can a First Amendment right to mediated seeing leave us with the ability to counter the dangers of distortion—and possibly manipulation—that come with such mediation? Does the right to share videos give individuals a right to send us a deepfake, which fabricates images with artificial intelligence, and then present them as genuine camera footage? In the United States, the Supreme Court has said that individuals' right to speak includes a right to speak falsely—except, that is, when the falsity is accompanied by certain harms. The plurality of the Court said that lying is protected except when speakers engage in harmful communications of a kind the law has traditionally recognized as outside the Constitution's protections, or in an extraordinary circumstance when the government needs to restrict the lying to achieve a "compelling government interest."[88] In the case of deepfakes that successfully deceive us, the technological mediation of our seeing or other perception enabled by others' sharing of video and audio does not do what it should ideally do. It does not, to use the Third Circuit's words, allow us to "see and hear more accurately."[89] Rather, it hijacks our confidence in video's accuracy to induce belief in images that it has disguised as real camera footage. In this kind of circumstance, there is a case to be made that First Amendment protection should give way: Here, one might argue that we do not need protection against government limits on regulation of videos. Rather, to see accurately, we may need the government's assistance in preventing the speakers who are sharing videos from substituting their own fabricated images for genuine camera records. Or perhaps, we should be able to claim a First Amendment freedom-of-thought right *against* whoever is using deepfake or other technology to distort our view of the world—whether it is a government official *or a private actor*. Or the solution may have to be partly technological rather than legal: We may need to counter technological distortions with other technologies that identify and perhaps correct them.

This challenge of deepfake distortion is perhaps most significant when the enhancement to our vision comes in the form of a video recording that counts as *someone else's expression*. After all, if a video they send is

---

[88] United States v. Alvarez, 567 U.S. 709, 719 (2012) (plurality opinion).

[89] Fields v. City of Philadelphia, 862 F.3d 353, 360 (3d Cir. 2017).

as much *their speech* as a lie they tell us, that would seem to indicate they are free to make the video a lie as well. I have argued in past work that—given the extent to which we have relied on video recordings in many contexts—a deepfake video should *not* be treated as equivalent to a statement of theirs. Except when it comes from a doctor, a lawyer, or someone else with whom we have a fiduciary relationship, a statement from someone else is something we should know we may have to treat with skepticism. The First Amendment, Justice Robert Jackson said, demands that each person serves as their own "watchman for truth."[90] But we cannot be expected to distrust *our perception* in the same way. And I argued in Chapter 2, even if recordings are embedded in *others'* communications, they are also enhancements of *our own* perceptions. Video recordings allow individuals to overcome the problem that arises from the *Zemel* court's recognition that they do not have a right to *physically be* any place that they have a strong interest in *seeing*. Recordings shared with us close the gap between where we have a right to freely exercise our powers of observation and where we would like to do so. They allow someone in the latter place to give us a sensory experience of being there. But if this is true, our interlocutors are to some extent like fiduciaries: We are relying on them to extend our powers of perception in ways that we cannot do so ourselves. It is only if those interlocutors act as transmitters of what is to be seen and heard at their location rather than as fabricators that video recording can, as the Third Circuit put it, let us "see and hear more accurately."

Of course, as I've written of deepfakes before, it is quite possible that some individuals might welcome them. The same technology that can be used to create videos that deceive individuals by mimicking reality can also be used to create impressive art.[91] As noted above, deepfakes are animations that happen to look almost indistinguishable from genuine camera footage. But individuals who know the deepfake is a fiction may be intrigued or entertained by the fictional scenario it portrays. It might, for example, allow a person to see a speech by Abraham Lincoln, Benjamin Disraeli, or Queen Victoria that looks as vivid as footage in a modern television broadcast or that depicts an alternate history not with actors but

---

[90] Thomas v. Collins, 323 U.S. 516, 545 (1945) (Jackson, J., concurring).
[91] *See* Marc Jonathan Blitz, *Lies, Line Drawing, and (Deep) Fake News*, 71 Okla. L. Rev. 59, 114 (2018).

by creating a deepfake that uses the images of real historical figures. Or individuals may wish to converse with people they know to be deepfakes of relatives who died long ago. As long as such deepfakes don't purport to be real—or, given what we know about the subjects they depict, have no chance of convincing us that they are real—then such videos might retain First Amendment protection that their deceptive counterparts ought to lack. In some very rare cases, an audience may even welcome the chance to be deceived (for example, in a game where the point is to correctly determine if a video is a fake or not).[92]

Of course, one challenge is how the law can differentiate between artistic and deceptive deepfakes. This is a challenge I have elsewhere offered a framework for addressing—arguing that a form of intermediate scrutiny akin to that used in *United States v. O'Brien* can provide a doctrinal tool for addressing the challenge: As Chapter 3 notes, the Court, in *O'Brien*, said that regulations that target the non-expressive component of O'Brien's draft card burning (that is, the physical destruction of the card) are subject to only moderate skepticism—intended to ensure that they don't impose substantially broader restriction on speech than is necessary—but regulations that *target* O'Brien's message are presumptively unconstitutional. Courts might similarly adopt a deferential posture toward regulations that target the deceptive components of a deepfake but show extreme skepticism toward regulations that target their message or the artistic elements in them.[93]

There is also another possible response that individual perceivers can make to guard against harms from deceivers: As I noted in Chapter 2, some means of extended perception are designed not to work through communication, but to free us from having to rely upon it: If the video we see comes to us not from a person but rather unaltered, from a machine, there is less possibility for someone to manipulate the sensory information we receive. Imagine, for example, that individuals can subscribe to a service that will send them camera footage, or perhaps a live feed, from numerous locations around the world. Global mapping services such as Google Streetview already provide visual information of this kind (albeit primarily with periodically updated still images of streets, houses, and

---

[92] *See* Marc Jonathan Blitz, *Deepfakes and Other Non-Testimonial Falsehoods: When Is Belief Manipulation (Not) First Amendment Speech?*, 23 Yale J. L. & Tech. 160, 257–261 (2020).

[93] *Id. at* 281–300.

other public structures rather than videos). Or imagine that such a service lets individuals have telepresence in various locations around the world where the company has installed cameras capable of sending immersive VR settings to a computer somewhere else in the world. To the extent the individual can trust the company that is providing such a service, it might serve as a trustworthy source against which they can test the veracity of videos posted by other individuals. Or imagine that a person uses a wearable computing device that can link to cameras at, and provide a sense of presence in, a multitude of different locations.

In fact, in an analysis of the legal implications of deepfakes, Bobby Chesney and Danielle Citron propose a mechanism like this for people to refute deepfakes that falsely depict *them*. A trustworthy, reliable, and secure lifelogging service, they say, may allow individuals to create continuous recordings of their life to refute falsification of it by deepfake creators. Specifically, they propose "immutable life logs or authentication trails that make it possible for a victim of a deepfake video to produce a certified alibi credibly proving that he or she did not do or say the thing depicted."[94] These, they add, would have to be created by a "strong reputation for the immutability and comprehensiveness of its data."[95]

In these passages, Chesney and Citron are mostly focused on how a lifelog service can address reputational harm. The lifelogger doesn't really need the lifelog themselves to know the deepfake is falsely portraying them. After all, they can compare the deepfake to their memories of their own experience. It is others who may be deceived by the fake video. The proposal above, by contrast, is designed to help those who want to avoid being fed false perceptions (not those defamed by such falsification). A service that sends me automated, unaltered recordings, secure live feeds, or reliable technologies for telepresence can do so. Of course, the danger of falsification or fabrication does not vanish entirely. There is, in these arrangements, still a company that serves as an intermediary—operating the cameras and ensuring they are protected against hacking. It is also possible that a hacker may succeed in altering extended perception anyway. However, this kind of solitary extended seeing nonetheless has certain advantages over one where the source of the video may be able to

---

[94] Bobby Chesney & Danielle Citron, *Deep Fakes: A Looming Challenge for Privacy, Democracy, and National Security*, 107 Cal. L. Rev. 1753, 1814 (2019). *See also* Mann, *supra* note 39 and text accompanying note 39.

[95] *Id.*

claim First Amendment rights to alter the video they send. One is that, unlike an individual speaker who posts a video, an intermediary playing such a role is not an entity that has its own First Amendment right to change the content of the video—any more than a cell phone company has a First Amendment right to fabricate the conversations that occur over its lines. To use the language used in many cases on whether intermediaries have First Amendment rights, a service that allows me to extend my perception in this way is acting as a "conduit" or channel of extended perception, not as a speaker itself. Alternatively, it might be considered a variant of what Jack Balkin has called an "information fiduciary" who is bound, in its management of information (perceptual inputs in this case), to act in accordance with a duty to its clients rather than in a way that furthers its interests.[96]

This analysis of deepfakes also provides a concrete illustration of a point that was made earlier. I noted in Chapter 4 and the beginning of this chapter that a kind of extended seeing that is not essential for our autonomy at one time may become essential at a later time. Some of the forms of extended seeing I have just discussed—the immutable lifelogs proposed by Chesney and Citron to show others' accurate images of one's life, the network of video feeds a person can tap into from anywhere on the earth to obtain accurate images for their own benefit—might seem at first like extensions of perception that aren't crucial for autonomy. After all, we have long been able to live in freedom and navigate the world without being able to tap into secure video feeds from locations throughout the world. In a world where deepfakes are plentiful, however, what was previously a luxury might become a crucial corrective.

---

[96] *See* Jack M. Balkin, *Information Fiduciaries and the First Amendment*, 49 U.C. Davis L. Rev. 1183, 1188 (2016). See also James Grimmelmann, *Speech Engines*, 98 Minn. L. Rev. 868, 904 (2014); Anuj C. Desai, *Regulating Social Media in the Free-Speech Ecosystem*, 73 Hastings L.J. 1481, 1484 (2022)(describing how courts have distinguished between "speaker[s]," who are protected by the First Amendment, and "'conduit[s]' for other people's speech" which generally are not).

Thinking About Rights, Technological Development, and Perceptual Enhancement: A Summary

This chapter began by noting that the case for a right to perceive our surroundings becomes more complex as one moves further away from the "core" areas where constitutional rights to perceive can be most clearly anchored: In a First Amendment right to record (and communicate our recordings) and in a right to personal autonomy that covers our use of our visual system. I have argued in this chapter that, as central as these rights are, they do not cover all of the ways that technologically enhanced seeing can be a core part of our cognitive liberty. That does not mean that a right to such conduct can be unlimited: It has to be defined in a way which leaves room for a right not only to see but also to remain unseen where we wish to retain our privacy. It also has to leave room for us to recruit other actors in society to ensure where see accurately and in a way that is consistent with, rather than undermining, our autonomy.

References

Michael Adler, *Cyberspace, General Searches, and Digital Contraband: The Fourth Amendment and the Net-Wide Search*, 105 Yale L.J. 1093 (1996).

Jack M. Balkin, *Information Fiduciaries and the First Amendment*, 49 U.C. Davis L. Rev. 1183, 1188 (2016).

Frank Biocca and Ben Delaney, *Immersive Virtual Reality Technology*, in ed. Frank Biocca and Mark R. Levy, *Communication in the Age of Virtual Reality* (1995).

Marc Jonathan Blitz, *A First Amendment for Second Life: What Virtual Worlds Mean for the Law of Video Games*, 11 Vand. J. Ent. & Tech. L. 779 (2009).

Marc Jonathan Blitz, *The First Amendment, Video Games, and Virtual Reality Training*, in ed. Woodrow Barfield and Marc Jonathan Blitz, *The Law of Virtual and Augmented Reality* (Edward Elgar 2018).

Marc Jonathan Blitz, *Lies, Line Drawing, and (Deep) Fake News*, 71 Okla. L. Rev. 59 (2018).

Marc Jonathan Blitz, *Deepfakes and Other Non-Testimonial Falsehoods: When Is Belief Manipulation (Not) First Amendment Speech?*, 23 Yale J. L. & Tech. 160 (2020).

Kate E. Bloch, Virtual Reality: Prospective Catalyst for Restorative Justice, 58 Am. Crim. L. Rev. 285 (2021).

Joseph Blocher, *Free Speech and Justified True Belief*, 133 Harv. L. Rev. 439 (2019).

Richard Glen Boire, *On Cognitive Liberty*, Part I, 1–3 Journal of Cognitive Liberties 1(1)(1999–2000).

Jan Christoph Bublitz, *A History of Cognitive Liberty*, in Jan Christoph Bublitz and Marc Jonathan Blitz, *The Law and Ethics of Freedom of Thought*, Volume II (forthcoming 2025).

Alan K. Chen and Justice Marceau, Truth and Transparency: Undercover Investigations: In Twenty-First Century 162 (Cambridge University Press 2023).

Bobby Chesney and Danielle Citron, *Deep Fakes: A Looming Challenge for Privacy, Democracy, and National Security*, 107 Cal. L. Rev. 1753, 1814 (2019).

Carolina Cruz-Neira, Daniel J. Sandin, Thomas A. DeFanti, Robert V. Kenyon and John C Hart (1 June 1992). "The CAVE: Audio Visual Experience Automatic Virtual Environment". Commun. ACM. 35 (6): 64–72.

Thomas Emerson, *A System of Freedom of Expression* (1970).

Nita A. Farahany, *The Battle for Your Brain: Defending the Right to Think Freely in the Age of Neurotechnology*, 214 (St. Martins' Press 2023).

Chris Frith, *Making Up the Mind: How the Brain Creates Our Mental World* (Blackwell Publishing 2007).

David S. Han, *Constitutional Rights and Technological Change*, 54 U.C. Davis L. Rev. 71 (2020).

John A. Humbach, *Privacy and the Right of Free Expression*, 11 First Amend. L. Rev. 16, 43–44 (2012).

Shiqi Jiang, et al., *Memento: An Emotion-Driven Lifelogging System with Wearables*, 15 ACM Trans. Sen. Netw. 1 (Jan. 2019).

Margot Kaminski, *Privacy and the Right to Record*, 97 B.U. L. Rev. 167 (2017).

Christoph Koch, *The Quest for Consciousness: A Neurobiological Approach* (Roberts & Co, 2004).

Myron W. Krueger, *Artificial Reality* II (1991).

Steve Mann, Jason Nolan, and Barry Wellman, *Sousveillance: Inventing and Using Wearable Computing Devices for Data Collection in Surveillance Environments*, Surveillance & Society1(3): 331–355 (2002).

Nancy Messieh, *Keeping a Lifelog: The Definitive Guide*, TNW, Jul. 11, 2011.

John Stuart Mill, *On Liberty and Other Essays* 17, 37 (John Gray ed., 1991).

Alva Noe, *Out of Our Heads: Why You Are Not Your Brain, and Other Lessons from the Biology of Consciousness* 139 (Hill and Wang 2009).

Robert Nozick, *Anarchy, State and Utopia* 42 (1974).

Thomas M. Scanlon, *Rights and Interests*, in eds. Kaushik Basu, and Ravi Kanbur, *Arguments for a Better World: Essays in Honor of Amartya Sen: Volume I: Ethics, Welfare, and Measurement* (Oxford, 2008).

Wrye Sententia, *Neuroethical Considerations: Cognitive Liberty and Converging Technologies for Improving Human Cognition*, Annals of the New York Academy of Sciences. 1013 (1) (2004).
Anil Seth, *Being You: A New Science of Consciousness* (Penguin Publishing 2021).
Jocelyn Simonson, *Copwatching*, 104 Cal. L. Rev. 391 (2016).
Scott Skinner-Thompson, *Recording as Heckling*, 108 Geo. L.J. 125 (2019).
Cass R. Sunstein, *Legal Reasoning and Political Conflict* 63(1996).
Ivan E. Sutherland, *The Ultimate Display*, 2 Proceedings of IFIP Congress 506–08 (1965), republished in Bruce Sterling, Augmented Reality: "The Ultimate Display" by Ivan Sutherland, 1965, Wired, Sep. 28, 2009, available at https://www.wired.com/2009/09/augmented-reality-the-ultimate-display-by-ivan-sutherland-1965/
Go Tanaka, et al, *GPS-Based Daily Context Recognition for Lifelog Generation Using Smartphone*, 6 Int'l J. of Adv. Comp. Sci. and Appl. 2 (2015).
John Vince, *Virtual Reality Systems* 14 (1995).
Christina E. Wells, Reinvigorating Autonomy: Freedom and Responsibility in the Supreme Court's First Amendment Jurisprudence, 32 Harv. C.R.-C.L. L. Rev. 159, 170 (1997).
Timothy Zick, *Clouds, Cameras, and Computers: The First Amendment and Networked Public Places*, 59 Fla. L. Rev. 1 (2007).

Cases

ACLU of Ill. v. Alvarez, 679 F.3d 583, 595 (7th Cir. 2012).
Askins v. U.S. Dep't of Homeland Sec., 899 F.3d 1035, 1043 (9th Cir. 2018).
Burstyn v. Wilson, 343 U.S. 495 (1952).
Brown v. Entm't Merchants Ass'n, 564 U.S. 786 (2011).
California v. Greenwood, 486 U.S. 35 (1988).
Camara v. Mun. Court of San Francisco, 387 U.S. 523, 535 (1967).
Carpenter v. United States, 585 U.S. 296 (2018).
Cruzan v. Director of Missouri Department of Health, 497 U.S. 261 (1990).
District of Columbia v. Heller, 554 U.S 570, 582 (2008).
Dow Chem. Co. v. United States, 476 U.S. 227, 238–239 (1986).
Fields v. City of Philadelphia, 862 F.3d 353 (3d Cir. 2017).
Fordyce v. City of Seattle, 55 F.3d 436, 439 (9th Cir. 1995).
Glik v. Cunniffe, 655 F.3d 78, 82 (1st Cir. 2011).
Irizarry v. Yehia, 38 F.4th 1282, 1289 (10th Cir. 2022).
Kyllo v. United States, 533 U.S. 27 (2001).
Leaders of a Beautiful Struggle v. Baltimore Police Dep't, 2 F.4th 330 (4th Cir. 2021).
New York v. Burger, 482 U.S. 691 (1987).
Packingham v. North Carolina, 582 U.S. 98 (2017).

Paris Adult Theatre I v. Slaton, 413 U.S. 49 (1973).
Porter v. Gore, 354 F. Supp. 3d 1162, 1171 (S.D. Cal. 2018).
Roth v. United States, 354 U.S. 476 (1957).
Smith v. City of Cumming, 212 F.3d 1332, 1333 (11th Cir. 2000).
Sorrell v. IMS Health Inc., 564 U.S. 552 (2011).
Terry v. Ohio, 392 U.S. 1, 8–9 (1968).
Thomas v. Collins, 323 U.S. 516, 545 (1945) (Jackson, J., concurring).
Turner v. Lieutenant Driver, 848 F.3d 678, 690 (5th Cir. 2017).
Va. Pharmacy Bd. v. Va. Consumer Council, 425 U.S. 748, 763 (1976).
United States v. Alvarez, 567 U.S. 709, 719 (2012) (plurality opinion).
United States v. Di Re, 332 U.S. 581 (1948).
United States v. Knotts, 460 U.S. 276, 282 (1983).
United States v. O'Brien, 391 U.S. 367 (1968).

INDEX

A
Artificial retina, 11, 116, 154
Augmented Reality (AR). *See* Extended reality (XR)
Autonomy, 12–14, 17, 23, 29, 31, 49, 50, 52, 54, 60, 90, 93, 95, 103, 104, 106, 109, 111, 117, 118, 123, 127, 128, 132, 133, 139, 140, 145–147, 152, 157, 158, 167, 172, 178, 179

B
Biology of vision
  photoreceptors, 114
  retina, 113, 114
  visual cortex, 114–116
Bionic eyes. *See* Artificial retina
Brain-computer interface (BCI), 11, 23, 30, 74, 91, 107, 110, 111, 116, 117, 127, 129, 149, 153, 155, 166

C
Cognitive liberty, 15, 16, 101, 102, 117, 149, 158–160, 165, 167, 168, 172, 179

D
Deepfake, 8, 33, 157, 174–178
Democratic deliberation, 142
Diaries, 88
Drones, 5, 7, 10, 12, 18, 20, 21, 29–31, 40, 50, 51, 53, 60, 65, 82, 94, 107, 124, 126, 129, 131, 138, 164, 169

E
Enhancement, 12, 128–130, 165, 170, 171, 174
Environment, access to (role in vision), 30, 38, 53, 114, 170
Extended mind, 23, 120–122, 124, 125

Extended reality (XR), 6, 8, 17, 59, 92, 107, 119, 130, 157, 165
  augmented reality (AR), 6, 17, 19, 92, 122, 154–155
  telepresence, 130
  virtual reality (VR), 6, 17, 92

F

First Amendment, 2, 4, 11–14, 16, 20–23, 28, 33–42, 44–50, 54, 58–70, 73, 75–79, 81–87, 89, 92–95, 100, 102, 105, 122, 124, 129, 130, 132, 138, 140–142, 146–148, 152, 157, 161–163, 165, 169, 172, 174–176, 178, 179
  counter-coverage, 168, 172
  coverage, 66, 81, 91
  freedom of the press, 93
  government purposes, 72, 73, 75, 78, 83
  O'Brien test, 69
  public Concern, matters of, 3, 35, 50, 87, 162–164
  public forum doctrine, 47
  receive information, rights to. *See* First Amendment
  record, right to. *See* First Amendment
  social practices, 85–90, 92
  speech creation. *See* First Amendment
Fourteenth Amendment, 13, 36, 52, 60, 104, 124, 133, 140
  bodily integrity, 12, 104, 109
  due process, 104, 123
Fourth Amendment, 7, 39, 52, 102, 103, 132, 142–146, 148, 169–173
  magnification, 169, 171
  pat down searches, 103
  thermal imaging, 31, 171

Freedom of the press. *See* First Amendment

H

Hacking, 111, 177

I

image capture, pervasive, 15
Implants, 11, 110, 117, 139
Interests, relationship to rights, 78

L

Lifelog, 8, 10, 61, 107, 166, 177, 178

M

Mapping (virtual globes), 72, 166

N

Neurointerventions, 106

P

Pacemakers, 110, 111, 117, 119, 123, 126
Photography, 2, 16, 72
Prediction (role in vision), 150
Privacy, 4, 5, 7, 24, 32, 39, 40, 43, 54, 65, 76–78, 92, 105, 107, 119, 121, 124, 128, 129, 131, 132, 142–145, 148, 163, 164, 166, 168–170, 172, 173, 179
Prosthetics, 52
Public forum doctrine. *See* First Amendment

R

Receive information, rights to. *See* First Amendment

Record, right to. *See* First Amendment
Rights, generally, 39, 108

S
Science fiction, 6, 9, 166
Scrutiny (applied by courts), 7, 48, 73
　intermediate scrutiny, 83, 131, 132, 173, 176
　strict scrutiny, 69, 79, 83
Searches. *See* Fourth Amendment
Smartphones, 6, 10, 108, 117, 118, 128, 168
Sousveillance, 21, 156
Speech Creation. *See* First Amendment
Streetlight effect, 58, 59
Surveillance, 7, 14, 21, 31, 39, 49, 132, 143, 169, 171, 172

T
Technological change (effect on rights), 140–142, 172
Telepresence. *See* Extended reality (XR)
Travel, right to, 36, 37

U
Unmanned aerial vehicle (UAV). *See* Drones

V
Video games, 50, 74, 92, 140, 146, 155, 160, 161
Virtual globes. *See* Mapping
Virtual reality (VR). *See* Extended reality (XR)

0700031895326